A Practical Guide to Clinical Virology
Second Edition

Edited by

L. R. Haaheim
Professor of Medical Microbiology, Department of Microbiology and Immunology, University of Bergen, Bergen, Norway

J. R. Pattison
Director of Research, Analysis and Information, Department of Health, London, UK

R. J. Whitley
Department of Pediatrics, The Children's Hospital, The University of Alabama at Birmingham, Birmingham, USA

JOHN WILEY & SONS, LTD

Telephone (+44) 1243 779777

First edition published 1989
Reprinted February 1993, November 1994

This book is based on *Håndbok i Klinisk Virologi* edited by Gunnar Haukenes and Lars R. Haaheim, 1983.

ISBN 0 471 91978 0 (World excluding Scandinavia and Finland)
ISBN 82 419 0038 4 (Scandinavia and Finland)

Email (for orders and customer service enquiries): cs-books@wiley.co.uk
Visit our Home Page on www.wiley.co.uk or www.wiley.com

Other Wiley Editorial Offices

John Wiley & Sons Inc., 605 Third Avenue, New York, NY 10158-0012, USA

Jossey-Bass, 989 Market Street, San Francisco, CA 94103-1741, USA

Wiley-VCH Verlag GmbH, Pappalallee 3, D-69469 Weinheim, Germany

John Wiley & Sons Australia Ltd, 33 Park Road, Milton, Queensland 4064, Australia

John Wiley & Sons (Asia) Ptd Ltd, 2 Clementi Loop #02-01, Jin Xing Distripark, Singapore 129809

John Wiley & Sons Canada Ltd, 22 Worcester Road, Etobicoke, Ontario, Canada M9W 1L1

British Library Cataloguing in Publication Data

A catalogue record for this book is available from the British Library

ISBN 0 470 84429 9 ppc
ISBN 0 471 95097 1 pbk

Typeset by Dobbie Typesetting Ltd, Tavistock, Devon
Printed and bound in Great Britain by Biddles Ltd, Guildford, Surrey
This book is printed on acid-free paper responsibly manufactured from sustainable forestry in which at least two trees are planted for each one used for paper production.

CONTENTS

THE TYPING POOL

CONTRIBUTORS

Dr Gabriel Ånestad, Department of Virology, National Institute of Public Health, Geitmyrsveien 75, N-0462 Oslo, Norway

Professor Birgitta Åsjö, Centre for Research in Virology, Department of Microbiology and Immunology, The Gade Institute, University of Bergen, PO Box 7800, N-5020 Bergen, Norway
Tel: +47 55 58 45 08; Fax: +47 55 58 45 12; E-mail: birgitta.asjo@vir.uib.no

Professor Bjarne Bjorvatn, Centre for International Health, University of Bergen, Armauer Hansen's Building, Haukeland Hospital, N-5021 Bergen, Norway
E-mail: bjarne.bjorvatn@cih.uib.no

Dr Anne-Lise Bruu, Mikrobiologisk laboratorium, Sykehuset: Vestfold HF, Postboks 2168, Postterminalen, 3103, Tønsberg, Norway

Professor Are B. Dalen, Institute of Cancer Research, University of Trondheim, Medisinsk Teknisk Centre, Norway
Tel: +47 22 04 22 86; Fax: +47 22 04 24 47; E-mail: gabriel.anestad@folkehelsa.no

Gary L. Davis M.D., Director, Division of Hepatology, Medical Director, Liver Transplantation, Baylor University Medical Center, Dallas, Texas, USA

Professor Miklos Degré, Institute of Medical Microbiology, Rikshospitalet University Hospital, 0027 Oslo, Norway
Tel: +47 23 07 11 00; Fax: +47 23 07 11 10; E-mail: degre@labmed.uio.no

Dr Yuri Ghendon, Research Institute for Viral Preparations, 1 Dubrovskaya Street 15, 109088 Moscow, Russian Federation
Fax: 7 095 274 5710

Professor Lars R. Haaheim, Department of Microbiology and Immunology, University of Bergen, Bergen High Technology Centre, POB 7800, N-5020 Bergen, Norway
E-mail: lars.haaheim@gades.uib.no

Dr Neil A. Halsey, Johns Hopkins University, Department of International Health, 615 N Wolfe Street, Baltimore, MD 21205-2103, USA

Professor Gunnar Haukenes, Centre for Research in Virology, Bergen High Technology Centre, University of Bergen, PO Box 7800, N-5020 Bergen, Norway
E-mail: gunnar.haukenes@vir.uib.no

Dr Gunnar Hoddevik, Department of Virology, National Institute of Public Health, Geitmyrsveien 75, N-0462 Oslo, Norway

Dr Elisabeth Kjeldsberg, Prof Dahls gate 47, N-0367 Oslo, Norway

Dr Jonathan A. McCullers, Department of Infectious Diseases, St Jude Children's Research Hospital, 332 N Lauderdale Street, Memphis, TN 38105-2794, USA
Tel: +1901 495 5164; Fax: +1901 495 3099; E-mail: jon.mccullers@stjude.org

Dr Ivar Ørstavik, Chief Medical Officer, Division of Infectious Disease Control, Norwegian Institute of Public Health, P.O. Box 4404 Nydalen, N-0403 Oslo, Norway
Tel: +47 22 04 22 85; Fax: +47 22 04 24 47
E-mail: ivar.orstavik@folkehelsa.no

Professor John S. Oxford, Academic Virology, Department of Medical Microbiology, St Bartholomew's and the Royal London School of Medicine and Dentistry, Turner Street, Whitechapel, London E1 2AD, UK
Tel: +44 (0)207 375 2498

Professor Sir John R. Pattison, Director of Research, Analysis and Information, Department of Health, Richmond House, 79 Whitehall, London SW1A 2NS, UK
E-mail: john.pattison@doh.gsi.go.uk

Dr George Shaw, The University of Alabama at Birmingham, Department of Medicine, Birmingham, AL 35294, USA

Dr Enok Tjøtta, Micro-Invent AS, Høyenhallsvingen 23, N-0667 Oslo, Norway
Tel: +47 22 26 54 90

Professor Terje Traavik, Institute of Medical Biology, Department of Virology, N-9037 University of Tromsø, Norway

Professor Richard J. Whitley, The University of Alabama at Birmingham, Department of Pediatrics, The Children's Hospital Ambulatory Care Center 616, 1600 7th Avenue South, Birmingham, AL 35294-0011, USA
Tel: 001 205 934 5316; Fax: 001 205 934 8559; E-mail: r.whitley@peds.uab.edu

Donna Wiger, MSc, The Norwegian Medicines Agency, Sven Oftedals vei 6, N-0950 Oslo, Norway
E-mail: Donna.Wiger@legemiddelverket.no

Drs Randi and Arnt Winsnes, The Norwegian Medicines Agency, Sven Oftedals vei 6, N-0950 Oslo, Norway
E-mail: randi.winsnes@legemiddelverket.no; arnt.wisnes@sensewave.com; winsnes@sensewave.com

PREFACE

Since its first edition in 1989* the science of virology has moved forwards at an impressive pace. Modern technology has unravelled many complex aspects of the genetics, structure and immunology of viruses, whereas the diagnosis and treatment of our most common viral diseases have not enjoyed a similar impressive development. However, recent years have given us several new antivirals, and it is hoped that we will also see new and better vaccines for general use, as well as better diagnostic tools.

In this pocket-sized handbook we have attempted to meet the need for condensed and readily accessible information about viruses as agents of human disease. We hope that this book will provide useful information for all health-care professionals, in particular practising physicians, medical and nursing students, interns and residents. We have included some new chapters on hepatitis and herpes viruses to this new edition, whereas the arboviruses chapter has been taken out.

The cartoons will hopefully entertain as well as provide a helpful visual image of some salient points.

The gestation period for this new edition was very long. Hopefully the offspring will please.

Bergen, London, Birmingham AL
March 2002

LARS R. HAAHEIM
Department of Microbiology and Immunology
University of Bergen
Bergen

JOHN R. PATTISON
Department of Health
Whitehall
London

RICHARD J. WHITLEY
Department of Pediatrics
The University of Alabama at Birmingham
Birmingham AL

*Edited by Haukenes G, Haaheim LRH, Pattison JR as a follow-up and extension of the Norwegian book *Håndbok i klinisk virologi*, Universitetsforlaget, Bergen, 1983.

HELLO FOLKS!

PREFACE TO 1ST EDITION

In this pocket-sized handbook we have attempted to meet the need for condensed and readily accessible information about viruses as agents of human disease. We have endeavoured to combine convenience with a concise but comprehensive account of medical virology. In order to achieve our aims we have broken with the traditional designs of textbooks and manuals. Thus all main chapters are constructed in the same way with respect to headings and the location of each subject within the chapter. The reader will for instance always find 'Epidemiology' at the bottom of the fifth page of a main chapter. In order to provide a brief overview a summary page containing an abbreviated form of the subsequent information is located at the beginning of each chapter. The cartoon drawings also break with convention. Perhaps they will not only amuse you but prove to be instructive and leave a visual image of some salient points.

The present book represents a development of a Norwegian book (Håndbok i klinisk virologi, Universitetsforlaget, Bergen, 1983) edited by two of us (GH and LRH). It is not a textbook but a guidebook, and we have therefore included four comprehensive textbooks for further reading as references.

We hope this book will provide useful information for all health-care professionals, in particular practising physicians, medical students, interns and residents. If the book convinces readers that clinical virology is part of practical everyday medicine, we will have succeeded in our aims.

Bergen and London
May 1989

GUNNAR HAUKENES
LARS R. HAAHEIM
Department of Microbiology and Immunology
University of Bergen
Bergen

JOHN R. PATTISON
Department of Medical Microbiology
University College and Middlesex School of Medicine
London

BREAKING THE CODE

ABBREVIATIONS

AIDS	Acquired immunodeficiency syndrome
Anti-HBc Anti-HBe Anti-HBs	Antibody against the hepatitis B virus core, e and surface antigens
Arboviruses	Arthropod-borne viruses
ARC	AIDS-related complex
ATL	Adult T-cell leukaemia/lymphoma
AZT	Azidothymidine
BL	Burkitt's lymphoma (EBV)
BKV	Strain of human polyoma virus
CE	California encephalitis (virus)
CF(T)	Complement fixation (test)
CJD	Creutzfeldt–Jakob disease
CMV	Cytomegalovirus
CSF	Cerebrospinal fluid
EBNA	EBV nuclear antigen
EBV	Epstein–Barr virus
ELISA	Enzyme-linked immunosorbent assay
F protein	Fusion protein
Fr.	French
Ger.	German
Gr.	Greek
H	Haemagglutinin
HAM	HTLV-associated myelopathy
HAV	Hepatitis A virus
HBcAG HBeAG HBsAG	Hepatitis B virus core, e and surface antigens, respectively
HBIG	Hepatitis B immunoglobulin
HBV	Hepatitis B virus
HCV	Hepatitis C virus
HDV	Hepatitis D (delta) virus
HEV	Hepatitis E virus

HFRS	Haemorrhagic fever with renal syndrome
HI(T)	Haemagglutination inhibition (test)
HIV	Human immunodeficiency virus
HPS	Hantavirus pulmonary syndrome
HPV	Human papilloma virus
HSV	Herpes simplex virus
HTLV	Human T-cell leukaemia virus
IF(T)	Immune fluorescence (test)
IgA, IgG, IgM	Immunoglobulins of the classes A, G and M, respectively
IL-2	Interleukin 2
JCV	Strain of human polyoma virus
Lat.	Latin
LCR	Ligase chain reaction
N	Neuraminidase
NANB	Non-A, non-B (hepatitis)
NE	Nephropathia epidemica
NP	Nucleoprotein
NPC	Nasopharyngeal carcinoma (EBV)
NT	Neutralization test
PCR	Polymerase chain reaction
PGL	Persistent generalized lymphadenopathy (HIV infection)
PHA	Passive (indirect) haemagglutination
PML	Progressive multifocal leukoencephalopathy (polyoma virus)
RIA	Radioimmunoassay
RIBA	Radioimmunoblot assay
RSV	Respiratory syncytial virus
RT-PCR	Reverse transcriptase polymerase chain reaction
SRH	Single radial haemolysis
SSPE	Subacute sclerosing panencephalitis (measles virus)
TBE	Tick-borne encephalitis (virus)
TSP	Tropical spastic paraparesis
URTI	Upper respiratory tract infection
VCA	Viral capsid antigen (EBV)
VZIG	Specific VZ-immunoglobulin
VZV	Varicella–zoster virus

REFERENCES FOR FURTHER READING

Collier L, Oxford J. *Human Virology*, 2nd edn. Oxford University Press, Oxford, 2000.

Knipe DM, Howley PM et al. (eds). *Field's Virology*, 4th edn. Lippincott Williams & Wilkins, Philadelphia, 2001.

Zuckerman AJ, Banatvala JE, Pattison JR (eds). *Principles and Practice of Clinical Virology*, 4th edn. John Wiley & Sons, Chichester, 1999.

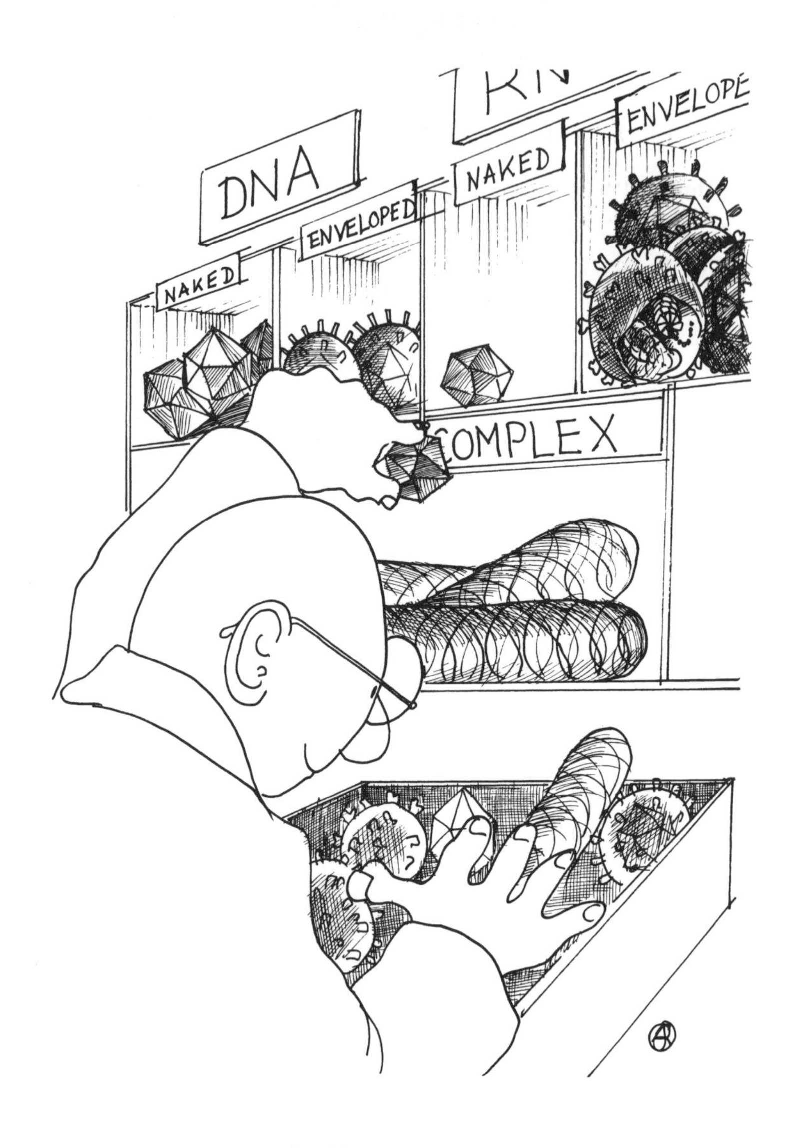

CLASSIFIED MATERIAL

1. CLASSIFICATION AND NOMENCLATURE OF HUMAN AND ANIMAL VIRUSES

Y. Ghendon

The present universal system for virus taxonomy includes family, genus and species. Virus families and subfamilies are designated by terms ending in *-viridae* and *-virinae*, respectively. Families represent clusters of genera of viruses with apparently common evolutionary origin. Genera are designated by terms ending in *-virus*. The criteria used for creating genera differ between families.

Virus characteristics used for classification vary from simple to complex structure, including nucleic acid and protein composition, virion morphology, strategy of replication, physical and chemical properties, etc.

More than 60 genera and about 25 families of human and animal viruses are recognized. Table 1.1 contains data on some families and genera of viruses infecting man.

Table 1.1 CLASSIFICATION OF HUMAN VIRUSES

Family Subfamily	Genus	Examples
Double-stranded DNA, enveloped virions		
Poxviridae		
Chordopoxvirinae	*Orthopoxvirus*	Smallpox (variola), vaccinia
	Parapoxviruses	Orf
	Molluscipoxvirus	Molluscum contagiosum viruses
	Yatapoxvirus	Yabapox virus, Tanapox virus
Herpesviridae		
Alphaherpesvirinae	*Simplex virus*	Herpes simplex virus 1 and 2
	Varicellovirus	Varicella-zoster virus
Betaherpesvirinae	*Cytomegalovirus*	Human cytomegalovirus
	Roseolovirus	Human herpesvirus 6
Gammaherpesvirinae	*Lymphocryptovirus*	Epstein–Barr virus
Double-stranded DNA, non-enveloped virions		
Adenoviridae	*Mastadenovirus*	Human adenoviruses
Papovaviridae	*Papillomavirus*	Human papillomavirus
	Polyomavirus	Human BK and JC virus

continued

Table 1.1 *continued*

Family Subfamily	Genus	Examples
Partial double-stranded partial single-stranded DNA, non-enveloped virions		
Hepadnaviridae	*Orthohepadnavirus*	Human hepatitis B virus
Single-stranded DNA, non-enveloped virions		
Parvoviridae		
Chordoparvovirinae	*Erythrovirus*	Parvovirus B19
Double-stranded RNA, non-enveloped virions		
Reoviridae	*Reovirus*	Reovirus types 1, 2, 3
	Rotavirus	Human rotaviruses (A and B)
	Orbivirus	Orungovirus, Kemerovo virus
	Coltivirus	Colorado tick fever virus
Single-stranded RNA, enveloped virions without DNA step in replication cycle		
(a) Positive-sense genome		
Togaviridae	*Alphavirus*	Sindbis virus (arbovirus group A)
	Rubivirus	Rubellavirus
Flaviviridae	*Flavivirus*	Yellow fever virus (arbovirus group B)
	Unnamed	Hepatitis C virus
Coronaviridae	*Coronavirus*	Human coronavirus
(b) Negative-sense, non-segmented genome		
Paramyxoviridae		
Paramyxovirinae	*Paramyxovirus*	Parainfluenzaviruses 1 and 3
	Morbillivirus	Measles virus
	Rubulavirus	Mumps virus, parainfluenzaviruses 2 and 4
Pneumovirinae	*Pneumovirus*	Respiratory syncytial virus
Rhabdoviridae	*Lyssavirus*	Rabies virus
	Vesiculovirus	Vesicular stomatitis virus
Filoviridae	*Filovirus*	Marburg and Ebola viruses
(c) Negative-sense, segmented genome		
Orthomyxoviridae	*Influenzavirus A, B*	Influenza A and B viruses
	Influenzavirus C	Influenza C virus
Bunyaviridae	*Bunyavirus*	Bunyamwera virus, La Crosse virus, California encephalitis virus
	Phlebovirus	Sandfly fever virus, Sicilian virus, Rift Valley fever virus, Uukuniemi virus
	Nairovirus	Crimean–Congo haemorrhagic fever virus
	Hantavirus	Hantaan virus, Seoul virus, Sin Nombre virus, Puumala virus

continued

Table 1.1 *continued*

Family Subfamily	Genus	Examples
Arenaviridae	*Arenavirus*	Lymphocytic choriomeningitis virus, Lassa virus, Venezuelan haemorrhagic fever virus
Single-stranded RNA, enveloped virions with DNA in the replication cycle		
Retroviridae	HTLV–BLV group	Human T-cell leukemia/ lymphotropic virus (HTLV-1 and HTLV-2)
	Spumavirus	Human foamy virus
	Lentivirus	Human immunodeficiency viruses (HIV-1 and HIV-2)
Single-stranded RNA, positive-sense, non-enveloped virions		
Picornaviridae	*Enterovirus*	Polioviruses 1–3, coxsackieviruses A1–22, A24, B1–6, echoviruses 1–7, 9, 11–27, 29–33, enteroviruses 68–71
	Hepatovirus	Hepatitis A virus
	Rhinovirus	Rhinoviruses 1–100
Caliciviridae	*Calicivirus*	Norwalk agent, hepatitis E virus?

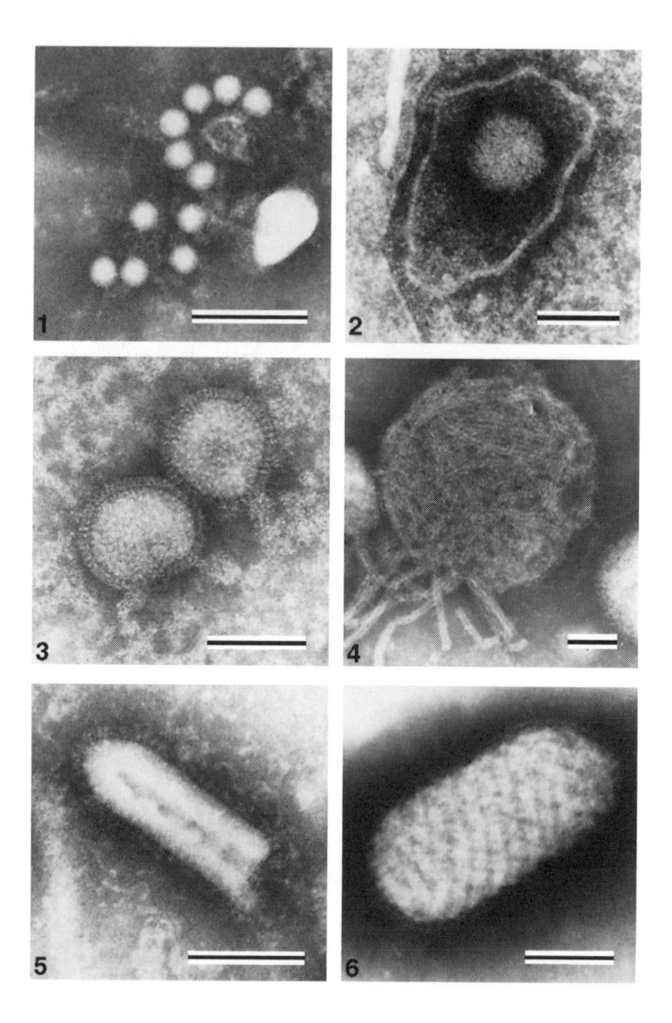

Figure 1.1 MORPHOLOGICAL FORMS OF VIRUSES: 1. poliovirus, naked RNA virus with cubic symmetry; 2. herpesvirus, enveloped DNA virus with cubic symmetry; 3. influenzavirus, enveloped RNA virus with helical symmetry; 4. mumps virus, enveloped RNA virus with helical symmetry—the helical nucleocapsid is being released; 5. vesicular stomatitis virus, morphologically similar to rabies virus; 6. orfvirus, also with a complex symmetry. Bars represent 100 nm (Electron micrographs courtesy of E. Kjeldsberg)

A LOAD OF TROUBLE

2. VIRUSES AND DISEASE

Virus = originally 'poisonous matter'.

G. Haukenes and J. R. Pattison

Viruses are the smallest known infectious agents. They are all built up of nucleic acid and protein coat(s) and may in addition have an outer lipoprotein envelope. They replicate in cells and may thereby lead directly to cell damage and cause disease. Alternatively, the host defences may lead to cell damage as they attempt to clear virus-infected cells.

TRANSMISSION/INCUBATION PERIOD/CLINICAL FEATURES

Virus infections are transmitted by inhalation, ingestion, inoculation, sexual contact or transplacentally. The incubation period differs greatly and may range from a few days (e.g. the common cold) to months (e.g. hepatitis B).

SYMPTOMS AND SIGNS

Systemic:	Malaise, Fatigue, Fever, Myalgia, Asthenia
Local:	Rash, Diarrhoea, Coryza, Cough, Lymphadenopathy, Neck Stiffness, Local Pain, Pareses, Conjunctivitis

Most infections are acute and of short duration. Some viruses become latent and may be reactivated, others are associated with persistent replication and chronic disease.

COMPLICATIONS

The infection may involve organs other than the one most frequently involved (e.g. orchitis in mumps). Complications may also result from immunopathological reactions (e.g. postinfectious encephalitis in measles) or from secondary bacterial infections (e.g. bacterial pneumonia in influenza).

THERAPY AND PROPHYLAXIS

A few antiviral drugs are available for clinical use in special therapeutic and prophylactic situations. Immunoglobulins and vaccines have been prepared for prophylaxis against a considerable number of virus infections.

LABORATORY DIAGNOSIS

Virus, viral antigen or viral genome may be detected in the early phase of acute disease by electron microscopy, immunological or molecular biological methods or virus isolation. Serologically the diagnosis can be made by demonstration of seroconversion, antibody titre rise or specific IgM.

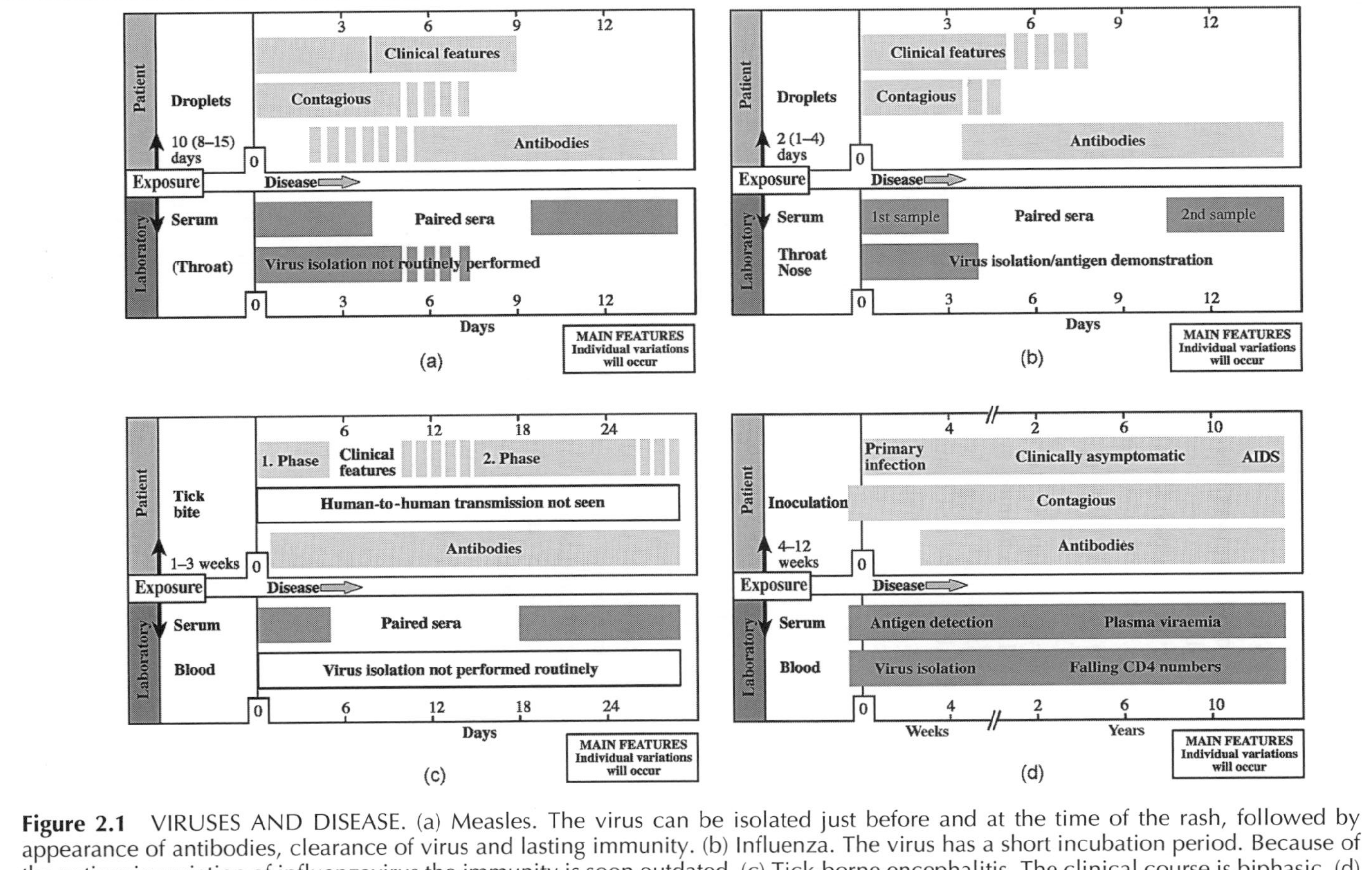

Figure 2.1 VIRUSES AND DISEASE. (a) Measles. The virus can be isolated just before and at the time of the rash, followed by appearance of antibodies, clearance of virus and lasting immunity. (b) Influenza. The virus has a short incubation period. Because of the antigenic variation of influenzavirus the immunity is soon outdated. (c) Tick-borne encephalitis. The clinical course is biphasic. (d) HIV infection. Infection has a prolonged course with persistence of the virus and antibody

CLINICAL FEATURES

SYMPTOMS AND SIGNS

Virus infections are mostly **transmitted** from acutely infected to susceptible individuals through the common routes: airborne, food, blood (inoculation) and direct contact. Some viruses infect the fetus (e.g. CMV, rubellavirus) and cause serious disease. Chronic and infectious carriers of virus are seen in hepatitis B, hepatitis C, hepatitis D and in AIDS virus infections. The **incubation period** may be a few days (upper respiratory infection, gastroenteritis), a few weeks (measles, rubella, mumps, varicella) or months (hepatitis, rabies, AIDS). **Prodromes** are commonly seen at the time when the virus spreads to the target organ (e.g. in measles, rubella and varicella). Local symptoms are due to the cell damage caused by virus replication in the target organ leading to inflammatory reactions (coryza, croup) or organ failure/dysfunction (icterus). Systemic symptoms (fever, malaise, myalgia) are secondary to release into the circulation of denatured and foreign protein from infected and degenerating cells. Some systemic symptoms (e.g. erythematous rashes) are immune mediated. Liberation of lymphokines from antigen-stimulated T-lymphocytes also contributes to the inflammatory response. Clinical **signs** are local inflammatory reactions such as oedema, hyperaemia and seromucous secretions, and general reactions such as leukocytosis or leukopenia with absolute or relative lymphocytosis. A polymorphonuclear leukocytosis is occasionally observed (e.g. in tick-borne encephalitis). Predominance of mononuclear cells is also found in the cerebrospinal fluid in meningitis. In acute uncomplicated cases the erythrocyte sedimentation rate and C-reactive protein values are within normal ranges, and the nitroblue tetrazolium test is usually negative unless there is extensive cell damage.

Differential diagnosis. It is of particular importance to exclude bacterial infections requiring antibacterial therapy, for example a purulent meningitis. Microbiological examinations may be required to establish the aetiological diagnosis.

CLINICAL COURSE

Most virus infections are acute and self-limiting, leading to lifelong immunity. Fulminant and lethal cases are usually the result of organ damage (poliomyelitis, hepatitis, encephalitis). Some infections have a biphasic clinical course (western tick-borne encephalitis, epidemic myalgia). Some viruses cause long-term infections. The pattern may be one of latency followed by reactivation and clinical recurrence (e.g. herpesviruses). Alternatively, there may be a persistent replication of virus but it may take years before clinical disease manifests itself (e.g. retroviruses and AIDS, hepatitis viruses and cirrhosis).

COMPLICATIONS

There is no clear distinction between that which is considered to be part of an unusually serious course and a complication. As a rule a complication is a manifestation of the spread of the infection to organs other than the most frequent targets (e.g. orchitis and meningoencephalitis in mumps) or a secondary bacterial infection (e.g. pneumococcal pneumonia following influenza). In some infections immunopathological reactions may lead to complications (e.g. postinfectious encephalitis in measles, polyarteritis nodosa in hepatitis B).

THE VIRUS AND THE HOST

The **virion** has a centrally located nucleic acid enclosed within a protein core or capsid. 'Naked' viruses are composed of this nucleocapsid only, while larger viruses have an envelope in addition. The nucleic acid is RNA or DNA, which is single- or double-stranded. If the RNA is infectious and functions as messenger RNA it is termed positive-stranded, otherwise minus-stranded (synonyms are positive- or negative-sense polarity). On the basis of the type of nucleic acid, the morphology of the capsid (cubical or helical) and the presence or absence of an envelope, a simplified scheme for classification can be constructed (see Chapter 1).

Since the cell cannot **replicate** RNA, viruses with an RNA genome furnish the cell with an RNA polymerase. The polymerase constitutes part of the core proteins of negative-stranded RNA viruses (e.g. influenzavirus), while positive-stranded RNA viruses (e.g. poliovirus) encode the production of the enzyme without incorporating it. Retroviruses have the enzyme reverse transcriptase which catalyses the formation of DNA from viral RNA; RNA is then synthesized from double-stranded DNA (provirus) by means of cellular enzymes. The viral envelope is a cell-derived lipid bilayer with inserted viral glycoproteins. The viral glycoproteins project from the surface of viruses and infected cells as spikes or peplomers and render the cell antigenically foreign, and as such a target for immune reactions.

The **pathogenesis** can in most cases be ascribed to degeneration and death of the infected cells. This may be mediated directly by the virus or by the immune clearance mechanisms. Denatured proteins elicit local inflammatory and systemic reactions. The local inflammatory response dominates the clinical picture in some infections, such as common colds, croup and bronchiolitis, while cell and organ failure or dysfunction is typical in poliomyelitis and hepatitis. Some infections are particularly dangerous to the fetus (CMV infection, rubella) or to the child in the perinatal period (herpes simplex, coxsackie B, varicella-zoster, hepatitis B and HIV infections). Bronchiolitis is seen only in the first 2 years of life, and croup mostly in children below school age. Otherwise the clinical course is not markedly different in children compared with adults.

The **defence mechanisms** involve phagocytosis, humoral and cellular immune responses and interferon production. In brief, interferon can arrest the local spread of the infection in the early phase; antibodies restrict further viraemic spread of the infection, mediate long-lasting immunity and sensitize infected cells for killing by macrophages and T-cells; while the cellular immune reactions include a series of events leading to the development of cytotoxic cells and release of lymphokines, including interferon. In the recovery from infection and protection against reinfection the various activities of the defence mechanisms are very interdependent.

EPIDEMIOLOGY

Human pathogenic viruses are maintained in nature mostly by continuous transmissions from infected humans or animals to susceptible ones. Chronic carriers of virus are important human reservoirs for hepatitis B and HIV and for some herpesviruses. Animal reservoirs play a role in rabies, in flavivirus and other 'arbovirus' infections, and in haemorrhagic fevers. Some virus infections lead to overt disease in most cases (measles, mumps, varicella, influenza) so the spread of the epidemic can easily be followed. In others clinical manifestations are exceptional (hepatitis B, enterovirus and CMV infections) and epidemic surveys may require laboratory tests. A balance is usually established for the maintenance of a virus in human populations. Antigenic changes (drift, shift) provide the underlying reason for epidemic spread of virus variants and subtypes (influenzavirus). A virus infection may be eradicated as with smallpox and, in the case of encephalitis lethargica (von Economo's disease), an infection appeared, existed for 10 years and then vanished.

THERAPY AND PROPHYLAXIS

Some progress has been made in the development of **antiviral drugs** in recent years. Main obstacles to a rapid breakthrough seem to be the rather late appearance of symptoms in relation to tissue damage and the potential cytotoxic effect of inhibitors of virus replication. Examples of antiviral drugs which are used clinically are aciclovir and trifluorothymidine in herpes simplex and varicella-zoster virus infections, azidothymidine in HIV infection and interferon in chronic active hepatitis B and C. Amantadine has proved effective in the prophylaxis of influenza A. Antivirals are dealt with in Chapter 4.

Immunoglobulins may provide short-term protection against certain virus infections. Normal human immunoglobulin is used in the prophylaxis of measles and hepatitis A, while specific immunoglobulins (produced from high-titred plasma) are needed for other infections (rabies, hepatitis B, varicella-zoster). A requirement for being effective is that the immunoglobulins are administered as early as possible after exposure, i.e. before the viraemic spread to the target organ.

Vaccines are now available against a number of virus infections. The vaccines are composed of either live attenuated virus (e.g. rubella, mumps, measles), inactivated whole virus (e.g. rabies, influenza) or viral components (e.g. influenza, hepatitis B). Second- and third-generation vaccines are recombinant DNA vaccines (hepatitis B) and synthetic peptide vaccines, respectively.

LABORATORY DIAGNOSIS

The aetiological diagnosis can be established by demonstration of virus, viral antigen or specific antibody. As a rule, virus or its antigens or genome can be demonstrated in the early acute phase of the disease, while antibodies appear from 5 to 20 days after exposure. Demonstration of virus in specimens taken from the affected organ is usually of diagnostic significance, although excretion of viruses (enterovirus, adenovirus) not associated with the disease concerned has to be considered. Demonstration of a concomitant antibody titre rise or of specific IgM may be valuable additional evidence. As many virus infections have an asymptomatic course, the mere demonstration of specific antibody is of limited value unless there is a titre rise or specific IgM is found. It may be difficult to distinguish reinfection or reactivation from a primary infection. Usually the IgM response is more marked in primary infections. All laboratory findings have to be evaluated in relation to the recorded time of exposure or onset of symptoms. It is important that the clinician gives adequate and relevant information to the laboratory. In return the laboratory will comment on the findings and advise regarding additional samples.

Laboratory testing is also performed in order to establish the immunity status of an individual. The methods used for screening (IgG tests) may be different from those used for establishing the diagnosis in acute infection (IgM test or paired sera examination). The clinician should therefore always state the clinical problem. Laboratory diagnosis is discussed in more detail in Chapter 3.

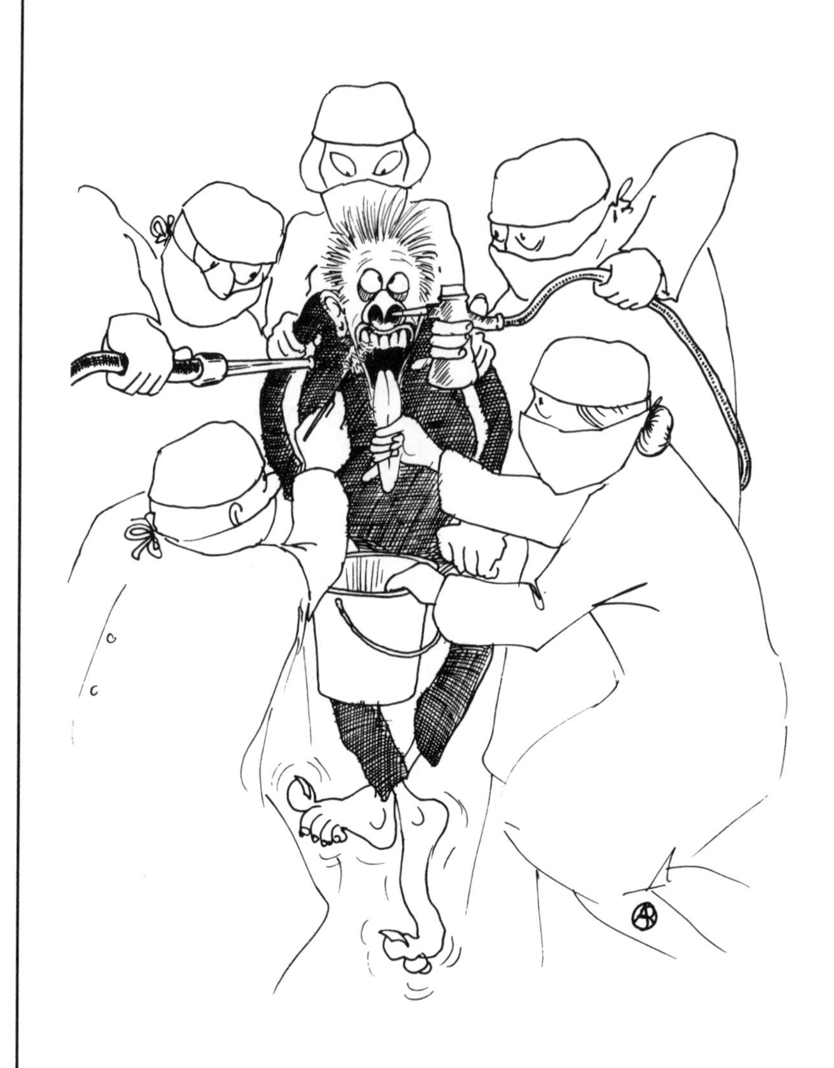

JUST TAKE A SAMPLE FOR VIRUS STUDIES

3. LABORATORY DIAGNOSIS OF VIRUS INFECTIONS

G. Haukenes and R. J. Whitley

Most virus infections run an asymptomatic course, or they are so mild that medical attention is not required. In many clinical cases an accurate aetiological diagnosis can be made solely on the basis of the clinical manifestations of the disease. Thus most cases of measles, varicella, zoster and mumps are diagnosed by the patient, his or her relatives, or by the family doctor. By contrast in other clinical situations, the resources required to establish an aetiological diagnosis are too great to justify virological examinations, for example in rhinovirus infections.

WHEN SHOULD VIROLOGICAL TESTING BE ORDERED?

In all clinical work the benefit of a precise diagnosis is indisputable. The consequences for the treatment of individual patients are obvious, and preventive measures can be taken to reduce the risk of transmitting the infection to others. In epidemics the laboratory diagnosis of a few early cases also benefits the doctor in that it allows confident aetiological diagnosis to be made for subsequent similar clinical cases. National and global epidemiological surveillance and control programmes will also require data from diagnostic laboratories. Decision as to the current composition of an influenza vaccine is one such example. The most common clinical situations requiring virological laboratory examinations are:

- **Respiratory infections**. Small children with severe respiratory illnesses and all age groups when influenza is suspected.
- **Gastroenteritis**. In general all cases which are severe and when there is an epidemic in progress.
- **Mumps**. In sporadic or doubtful cases, and in cases of orchitis, meningoencephalitis or pancreatitis when the clinical diagnosis of mumps is not certain. Immunity status screening for vaccination of adult or prepubertal males.
- **Rubella**. When rubella is suspected in a pregnant woman or in her family contacts. The immunity status of a woman should always be established in connection with premarital or family planning consultations and on the first consultation in her pregnancy. All cases of suspected congenital rubella require laboratory confirmation.

- **Measles**. In clinically doubtful cases, when SSPE is suspected and in cases of postinfectious encephalitis of unknown cause.
- **Varicella**. When the rash is not typical. Immunity status should be established in children before treatment with cytotoxic drugs and in women exposed to varicella in the last trimester of pregnancy.
- **Zoster**. Verification of the clinical diagnosis may be desirable, also for selection of donors of blood for preparation of hyperimmunoglobulin.
- **Herpes simplex**. In pregnancy, especially when genital herpes is suspected before delivery. In severe herpes simplex, generalized herpesvirus infection in newborn infants and cases of encephalitis.
- **Cytomegalovirus infections**. Screening of blood donors and of donors and recipients of tissues and organs. Cases of prolonged fever or mononucleosis-like disorders when heterophile antibody and anti-EBV tests are negative, especially if occurring during pregnancy or as part of a post-transfusion syndrome. Prolonged fever of unknown cause. Fever and pneumonia in immunocompromised individuals.
- **Epstein–Barr virus (EBV) infections**. When infectious mononucleosis is suspected and the diagnosis has not been made by tests for heterophile antibodies.
- **Hepatitis**. All cases of hepatitis should be examined for viral antigen and/or antibody. High-risk groups are screened for the chronic carrier state of hepatitis B and C viruses. Blood and tissue donors must be screened for HBsAg and anti-HBc and for anti-HCV. Immunity status is determined in high-prevalence or high-risk groups before vaccination against hepatitis B, or before vaccination or the repeated use of normal immunoglobulin to prevent hepatitis A.
- **Erythema infectiosum**. The clinical diagnosis may be uncertain, especially in non-epidemic periods. When parvovirus B19 infection is suspected in pregnancy. Cases of arthralgia.
- **Meningitis, encephalitis** and other severe disorders of the nervous system require microbiological and serological examinations to establish the aetiology.
- **HIV infection**. The clinical manifestations in any phase of an HIV infection comprise a wide range of syndromes, which will require testing for anti-HIV. Subjects at risk of contracting HIV infection have to be examined in accordance with national control programmes. All donors of blood and tissue (including breast milk) should be tested for anti-HIV.
- **HTLV infection**. Cases of T-cell leukaemia and progressive spastic paraparesis of unknown cause in individuals who may be at risk of exposure to HTLV-1. Screening of blood and tissue/organ donors for anti-HTLV-1/2 in accordance with national control programmes.

LABORATORY DIAGNOSIS

Virological diagnosis is based either on demonstration of the virus or its components (antigens or genome) or on demonstration of a specific antibody response. In some infections antibodies are detectable at the onset of clinical disease (e.g. poliomyelitis, hepatitis B (anti-HBc)), or the antibody appearance may be delayed by days (rubella), weeks or months (hepatitis C, HIV infection). Whenever an early diagnosis is important for the institution of antiviral therapy or some other interference measures, the possible use of methods that demonstrate the virus should be considered.

The **virus** can be demonstrated directly by electron microscopy (gastroenteritis viruses, orfvirus). Alternatively, infectious virus may be demonstrated after inoculation of cell cultures (enteroviruses, adenoviruses, herpes simplex virus, cytomegalovirus), embryonated eggs (influenzaviruses) or laboratory animals (coxsackievirus). Clinicians should carefully follow the instructions issued by their local laboratories with regard to sampling and transportation, especially if infectivity has to be maintained.

Viral genomes can be demonstrated by various nucleic acid hybridization techniques, either *in situ* or in tissue extracts (slot blot, Southern blot, *in situ* hybridization) using labelled DNA or RNA probes, or by methods that include amplification of the viral nucleic acid such as polymerase chain reaction (PCR) and ligase chain reaction (LCR). Both PCR and LCR are extremely sensitive, requiring strict precautions in the laboratory to avoid contamination. The gene technology methods are of particular importance for rapid diagnosis of infections that are accessible to antiviral treatment (herpes simplex encephalitis, CMV infection), for diagnosis of infection with viruses that cannot be cultivated (human papillomaviruses) or viruses that grow slowly in culture (enteroviruses), as well as in clinical situations where a definite diagnosis cannot be made by other means (possible HIV infection and hepatitis B or C in newborns and infants).

Several **virus antigen tests** are available for rapid diagnosis of virus infections. Methods most commonly used are immunofluorescence or immunoperoxidase for respiratory viruses, ELISA for HBsAg, HIV and rotavirus, latex agglutination for rotavirus, and reverse passive haemagglutination for HBsAg. Immunofluorescence and immunoperoxidase procedures depend on the sampling and preservation of infected cells, requiring rapid transport of cooled material. Alternatively, preparation of the slide has to be made locally. Blood (serum) and faeces can be sent in the usual way.

Antibody examinations are mostly performed with serum. Anticoagulants added to whole blood may interfere with complement activity and enzyme functions, and should be avoided. In certain situations (SSPE, herpes simplex encephalitis) antibody titration is performed on cerebrospinal fluid. Acute infection is diagnosed by demonstrating a rise in titre, seroconversion or specific IgM (or IgA). A rise in titre may be seen both in primary infections and in reinfection or after reactivation. A positive IgM test usually indicates a primary infection, but lower concentrations of specific IgM are found in reactivations (CMV infections and zoster) and reinfections (rubella). A variety of methods (complement fixation (CF), haemagglutination inhibition (HI), enzyme-linked immunosorbent assay (ELISA), immunofluorescence (IF)) are available for demonstration of antibodies, and the choice of test will depend on the virus and whether the clinical problem is the immune status or diagnosing an acute infection. Blood samples for demonstration of seroconversion or titre rise (paired sera) are taken 1–3 weeks apart, depending on the time of exposure or onset of symptoms.

INTERPRETATION OF RESULTS

To achieve the full benefit of virological tests, appropriate specimens must be taken at the optimum time and must be transported to the laboratory as recommended. In the laboratory, the virologist will decide on appropriate tests on the basis of the information given by the clinician. This information will also be important for the interpretation of the laboratory findings. Thus, a meaningful laboratory service depends on collaboration between the clinician and the virologist.

Isolation of a virus does not prove that the virus is the cause of the clinical condition concerned. Enteroviruses, for example, may be shed into the pharynx and the intestines for long periods after an acute episode. A concomitant antibody titre rise supports the evidence of a causal connection. By contrast, isolation of a virus from the blood or from the cerebrospinal fluid will usually be diagnostic whatever the antibody findings.

The mere demonstration of a high antibody titre is of limited diagnostic value and will have to be evaluated in relation to the clinical problem. The virologist will know the time after exposure or onset of symptoms that antibodies are detectable, when an antibody titre rise is expected, and for how long it may be possible to demonstrate specific IgM. It is therefore of crucial importance that the clinician provides the relevant data about time of possible exposure and onset of symptoms, and, in some clinical situations, information about pregnancy and vaccinations. The virologist can then comment on the findings and advise further tests if indicated.

THE CONSTRUCTION COMPANY

4. ANTIVIRAL DRUGS

J. S. Oxford and R. J. Whitley

The history of antiviral chemotherapy as a science is short, commencing in the 1950s with the discovery of methisazone which is a thiosemicarbazone drug inhibiting the replication of poxviruses. The experience of clinical application of antivirals is even shorter and most comprehensively involves 24 important licensed drugs: amantadine and the related molecule rimantadine, primary amines which inhibit influenza A viruses; the newer antineuraminidase drugs which inhibit both influenza A and B viruses; aciclovir and related acyclic nucleoside analogues inhibiting herpes type I and type II; zidovudine and the group of dideoxynucleoside analogues, non-nucleoside inhibitors of HIV reverse transcriptase and also inhibitors of the viral protease enzyme. Very extensive use has been made of aciclovir and the anti-HIV molecules, bringing wide recognition to the science of antivirals. In fact the most striking example of the potential of antivirals was the discovery and clinical application of zidovudine, within 2–3 years of the first isolation of the HIV-1 itself, and the more recent discovery and use in the clinic of additional antiretroviral drugs. As a comparison, and after a further 15 years of hard work, effective vaccines against HIV have yet to be developed.

But, of equal importance to the search for new inhibitors, is the attention to strategies to use existing compounds sensibly and to maximum clinical effect without squandering the discoveries. Viruses, particularly RNA viruses, can mutate rapidly and thus drug resistance to viruses could quickly become the major problem it already is with antibiotics and bacteria. Antiviral chemotherapists have already benefited from the clinical experience of preventing drug resistance against mycobacteria by using three drugs concurrently and combinations of two or three antivirals are now being used successfully to prolong the life of AIDS patients.

In the present chapter we will outline some of the underlying principles of antiviral chemotherapy, place emphasis on the most important existing licensed drugs and attempt a short stargaze into the future. Unfortunately the future may look a little bleak. History has come full circle and chemotherapists are now actively searching for new drugs against smallpox virus.

THE TARGET VIRUSES

HIV-1 is, and will probably remain, the prime focus of attention for antiviral chemotherapists for two reasons, namely medical and economic (Table 4.1). As regards the latter, it should be appreciated that a minimal cost of a drug development is $0.5 billion. A pharmaceutical company will not develop drugs

Table 4.1 FEATURES OF THE TARGET VIRUSES FOR CHEMOTHERAPY

Virus	Why are further antivirals required?	Potential problems
HIV	No vaccine exists. Retrovir and the other dideoxynucleoside analogues, non-nucleoside inhibitors and also protease inhibitors have only limited efficacy. The virus is worldwide and spreading rapidly in Asia and drug-resistant viruses are emerging	Drug resistance
Influenza	Epidemics occur yearly resulting in serious morbidity and death in 'at risk' groups. The vaccine is not 100% effective. Periodically worldwide pandemics sweep the world. Two new anti-NA drugs have been licensed recently to join the M2 blocker amantadine (Lysovir)	Drug resistance
Human herpes viruses (HHV1–8)	No vaccines exist. The disease is lifelong and recurrent infections are common. Prodrugs are now utilized but successful chemotherapy is restricted to one member of this large family, HSV-1	None
Respiratory viruses	A myriad of 150 common cold viruses, six adenoviruses, four parainfluenzaviruses and coronaviruses inhabit the upper respiratory tract. They may trigger serious bacterial infections or attacks of bronchitis	Impossible to differentiate clinically and hence a broad spectrum antiviral will be required
Hepatitis B and C viruses	Very common infections in many areas of the world. Interferon α is used in the clinic, as is lamivudine (3TC) and famciclovir against hepatitis B	Persistent chronic infection
Human papilloma (wart) viruses	Common virus which can be spread sexually	None envisaged
Smallpox	The threat of reemergence of monkey pox and camel pox or the use of bioterrorism	

against rare viral diseases. Front-line target viruses are therefore HIV-1, herpes, influenza and common cold viruses, with more recent attention on hepatitis B, hepatitis C and papilloma viruses. There is a further important fact which will encourage chemists to produce even more antivirals. A quasi-species RNA virus such as HIV existing as a 'swarm' of countless genetic variants will easily, by mutation and selection, evade the blocking effects of a single inhibitor. Therefore, as with tuberculosis, the practical answer is to find inhibitors of a wide range of virus-specific enzymes or proteins and to use them in a patient simultaneously. This search for new drugs will be a continuing need as it is with antibacterials. Similarly, inhibitors of pandemic and epidemic influenza A viruses will need the continuing attention of antiviral chemotherapists. The human herpes viruses (HHV1–8) cause a remarkably diverse range of important diseases and will continue to remain important targets, especially VZV (varicella-zoster virus or shingles), which will reach new importance in a world population with increasing longevity. Common cold viruses and other viruses of the respiratory tract cause pathogenesis in the upper respiratory tract during all months of the year in all countries of the world and hence have economic importance. The eight or so hepatitis viruses, and especially hepatitis B and C, are increasingly recognized as virus diseases where chronic or prolonged infection gives extensive opportunity to the application of therapeutic drugs. Papilloma viruses are extremely common, are considered to be oncogenic and can be spread sexually. Sadly we have now to add smallpox to our list as a possible bioterrorist virus. But there is another lesson to be learnt here: there are other pox viruses from monkeys and camels which cause disease in humans and they could emerge naturally, and in fact, are a bigger threat to our safety than a deliberate release.

There is therefore no shortage of viral targets for new drugs. The main problem, as ever, is the actual discovery of a novel drug. It must be clearly recognized that all the antivirals yet discovered have an extraordinarily restricted antiviral spectrum. For example, amantadine inhibits influenza A, but not influenza B virus, whilst aciclovir is highly effective against herpes simplex type I but has little or no effect against the herpes cytomegalovirus. Similarly the neuraminidase inhibitors only target influenza A and B viruses and have no effect against viruses of other families such as paramyxoviruses.

HOW ARE NEW ANTIVIRALS DISCOVERED?

To the present day our antivirals have been found by true Pasteurian logic, to be paraphrased as 'discovery favours with prepared mind'. In practical laboratory terms 'off-the-shelf' chemicals are subjected to a biological screen. A virus-susceptible cell line is incubated with a non-toxic concentration of novel drug and the 'target' or 'challenge' virus is then added. If the cell is rendered uninfectable or if there is a 10–100-fold reduction in the quantity of virions produced by the drug-treated cell, the drug is further investigated. The many stages of the lifecycle of a virus give chemotherapeutists the opportunity to design or find compounds which interrupt virion binding, penetration or more usually some vital step dependent upon a unique viral enzyme such as RNA polymerase, protease or integrase (Table 4.2). Virologists have screened through libraries of millions of already synthesized compounds, either using biological or, increasingly, automated ELISA screens against particular viral

Table 4.2 STEPS IN VIRUS REPLICATION THAT ARE SUSCEPTIBLE TO INHIBITORS

Target	Antiviral	Virus/infection
(1) Virus adsorption	Dextran sulphate	HIV-1
	CD4 (receptor)	HIV-1
(2) Viral penetration and uncoating	Amantadine (Symmetrel or Lysovir)*	Influenza A
	Rimantadine*	
	gp41 peptides (fusion)	HIV-1
(3) Virus-induced enzymes Reverse transcriptase	Zidovudine (AZT)	HIV-1
	Zalcitabine (ddC)	
	Didanosine (ddI)	
	Stavudine (D4T)	
	Lamivudine (3TC)	
	Delavirdine	
	Nevirapine	
	Efavirenz	
DNA polymerase	Aciclovir (ACV)*	Generalized herpes and shingles infections and genital HSV infections
	Penciclovir*	
	Ganciclovir*	Cytomegalovirus infections (e.g. pneumonia)
	Trifluorothymidine* (TFT)	Eye infections with HSV
	Foscarnet	CMV infections
	Cidofovir	Pox viruses
Protease	Saquinavir*	HIV
	Indinavir*	
	Nefinavir*	
	Ritonavir*	
Neuraminidase	Zanamivir*	Influenza A and B
	Oseltamivir*	
Viral protein synthesis	Interferon*	Many viruses
Free virus particle	Pleconaril	Rhinoviruses

*A licensed antiviral.

proteins. Once a molecule binding to a viral protein has been located, a more efficient molecule can be 'designed' by the chemists. Excellent examples of semi-designed antivirals are inhibitors of the common cold virus, which bind tightly to the viral capsid protein and which can be visualized by X-ray crystallography in the binding pocket on the virion surface, and also inhibitors of the influenzavirus neuraminidase enzyme. In the latter case the enzyme-active site had been identified as a saucer-like depression on the top of the viral neuraminidase protein and X-ray crystallography identified exactly which amino acids of the viral protein were interacting with an inhibitor. Chemists have modified an already discovered drug by addition of a single side chain to enable it to bind more strongly to the influenza neuraminidase protein and hence cause stronger inhibition of viral replication. The effects occur at a late stage of viral growth where the function of the neuraminidase is to cause release of newly synthesized virus from the infected cell.

But, do not forget that the members of the plant kingdom are excellent chemists as well and laboratories are intensifying the search to discover novel molecules in plant extracts which, by chance, inhibit viruses. There is a strong history here to remember, with A. Fleming's discovery of the penicillin antibiotic, synthesized by a penicillium mould on an orange.

Chemists also believe that thousands of nucleoside analogues remain to be synthesized and tested as antivirals. Alternatively, these compounds may already be in existence and on the shelf as part of a completely unrelated biological screening programme. The now classic anti-herpes nucleoside analogue aciclovir was initially synthesized as an anti-cancer drug. Aciclovir is structurally related to the natural nucleoside 2′ deoxyguanosine but has a disrupted sugar ring (acyclic). Nucleoside analogues are some of our most powerful antivirals and even more surprisingly some, like aciclovir, appear to

Table 4.3 RELATIVELY FEW POINT MUTATIONS IN VIRAL GENES CAN LEAD TO DRUG RESISTANCE

Virus	Drug	Specific mutations responsible for resistance
Influenza A	Amantadine, zanamivir and oseltamivir	Mutations in M2 gene (amantadine) and possibly in HA gene and NA genes (oseltamivir and zanamivir). Fortunately NA mutants are less able to spread
Herpes simplex	Aciclovir	Mutations in the viral DNA polymerase or TK enzyme. Importantly, some viruses without thymidine kinase or possessing an altered TK may be less virulent *in vivo* and so will not spread
HIV-1	Zidovudine and other dideoxynucleoside analogues	Five mutations in the reverse transcriptase gene
	Protease inhibitors	Mutations in the HIV protease

be extraordinarily safe in the clinic and, not surprisingly, have become the virologists' favourite molecule.

HOW IMPORTANT A PROBLEM IS DRUG RESISTANCE?

It must be acknowledged that the high mutation rates of the classic RNA viruses such as influenza and HIV will always result in 'resistance' problems for antiviral drugs. With influenza A virus a single mutation in the target M2 gene allows the mutated virus to escape from the inhibitory effects of amantadine (Table 4.3). Similarly with HIV-1, amino acid changes in the target viral reverse transcriptase enzyme allow the virus to replicate in the presence of zidovudine and other dideoxynucleoside analogues. Recent studies have shown that drug-resistant HIV mutants emerge within days of initiation of treatment of an infected patient with certain dideoxynucleoside analogues. This has led to the use of combination chemotherapy (highly active antiretroviral therapy (HAART)) using three inhibitors, two against the RT enzyme and one against the protease enzyme. With a DNA virus such as herpes the drug resistance problem is correspondingly less acute because of the lower virus mutation rates. There are proofreading enzymes already in the cell to correct errors in DNA-to-DNA transcription, but not to correct RNA-to-DNA or RNA-to-RNA molecular events. The first clinical trial in AIDS patients established that the mortality following administration of zidovudine alone was 17%, whereas if a patient was also administered didanosine or zalcitabine the mortality dropped to 10 and 12%, respectively. Addition of protease inhibitors and non-nucleoside inhibitors adds further benefits. These extra clinical benefits are assumed to accrue partly because of avoidance of drug resistance. Furthermore, DNA polymerase enzymes have a higher fidelity of reading and fewer transcription mistakes occur. However, although mutation rates may be 1000-fold less with herpesvirus than with an RNA virus, drug resistance does occur in immunosuppressed patients undergoing transplantation surgery or in AIDS patients where herpesviruses with mutations in the thymidine kinase gene allow the mutant to escape from the effects of aciclovir. So in immunocompromised patients the exceedingly rare mutated virus can emerge and dominate the virus population in the patient.

The experience of clinical bacteriologists treating infections with *Mycobacteria tuberculosis* has led to the use of combinations of two or three drugs with different points of action. For example, if the chance of a zidovudine-resistant mutant of HIV occurring is 10^3, then by treating a patient with three antiviral drugs at the same time the chance of a mutant arising with simultaneous genetic changes at all three critical viral sites would exceed 10^9 (but with a different target for each) and would therefore be vanishingly small. We have therefore entered the time of combination chemotherapy for viruses, particularly HIV.

HOW ARE ANTIVIRALS USED IN CLINICAL PRACTICE?

Antivirals can be used prophylactically to prevent a virus infection in a person who has not yet been infected but who will be in contact with others who are infected, in much the same way as most vaccines are used. Prophylactic protection with a chemical antiviral is more rapid in onset than that induced by vaccines, since some antiviral protection would be anticipated within 30 minutes of drug administration. Most antivirals are administered by mouth and rapid adsorption of drug and spread to all tissues of the body is often achieved. To maintain active levels of drug in the target organ redosing is required, perhaps twice daily. Clinical use is less convenient if oral absorption is poor, and if the drug has to be given by intravenous infusion or, in the case of respiratory infections, by aerosol or nasal spray. However, it could be argued that with respiratory infection direct application of a drug to the nose and airways could have medical advantages.

Two anti-herpes prodrugs and more recently an anti-influenza neuraminidase inhibitor have been studied and licensed, which are inactive themselves but are converted by enzymes in the patient to the active antiviral (prodrugs).

It should be remembered that when drug prophylaxis is discontinued the patient becomes susceptible to virus infection unless a subclinical infection has occurred in the first instance, giving the patient some immunity to reinfection with the same virus.

An example of prophylaxis is the use of amantadine or the neuraminidase inhibitors to prevent spread of influenza A virus within families. The drugs are given to the family members who are still well, but in contact with an ill person perhaps in their own family. Under these circumstances nearly 90% protection can be achieved.

Most often antivirals are used therapeutically, being administered either after infection or even after the first clinical signs of the disease are noted. In this situation further progression of the disease may stop and/or the virus infection may resolve more rapidly. Therapy is the favoured mode with HIV-1-infected patients, although the drug may be given before overt clinical signs of disease both to delay the time before early symptoms occur and also to lessen the chance of drug resistance occurring. There is still active debate about when to initiate chemotherapy, early in the disease or later. Aciclovir may be used to prevent recurrent herpes infections and this is a therapeutic use because the patient is already infected with the virus. The three anti-influenza drugs can all be used to ameliorate clinical symptoms and also to prevent secondary complications of influenza, but have to be given to the patient within 48 hours of onset of symptoms. It should be appreciated that even experienced clinicians are still on a learning curve with the prevention and treatment of most viral infections using antivirals.

WHICH ARE THE CLINICALLY EFFECTIVE ANTIVIRALS?

A short list of antivirals in order of their clinical usefulness and effectiveness would be headed by aciclovir, followed by the dideoxynucleoside analogues and the new neuraminidase inhibitors and, finally, amantadine (Figures 4.1–4.3). A range of retroviral agents are now used against HIV-1 and several of

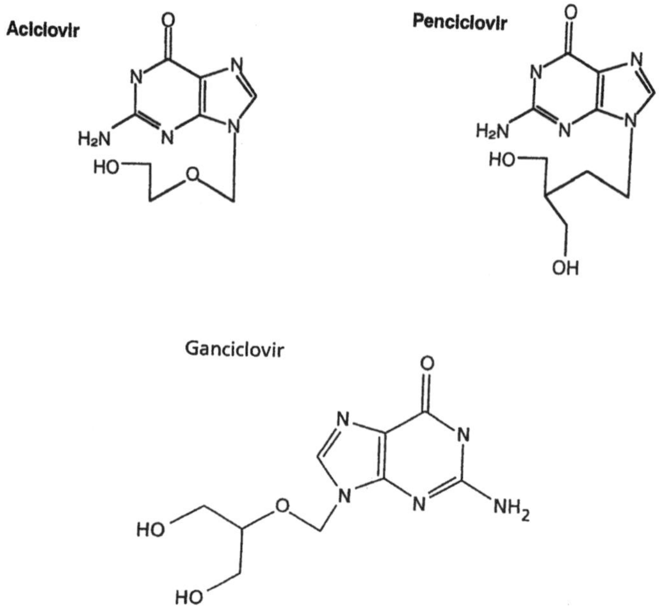

Aciclovir

Penciclovir

Ganciclovir

Figure 4.1 CLINICALLY EFFECTIVE DRUGS AGAINST HERPESVIRUSES

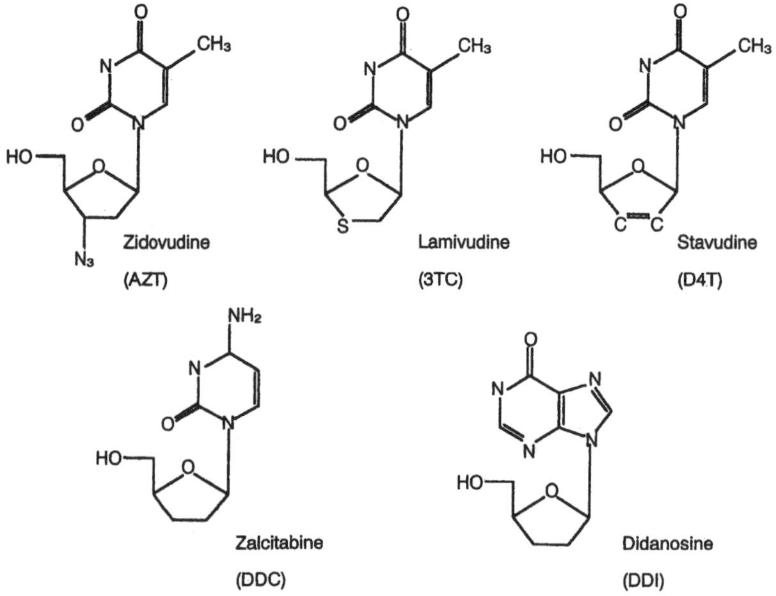

Zidovudine (AZT)

Lamivudine (3TC)

Stavudine (D4T)

Zalcitabine (DDC)

Didanosine (DDI)

Figure 4.2 CLINICALLY EFFECTIVE DRUGS AGAINST HIV

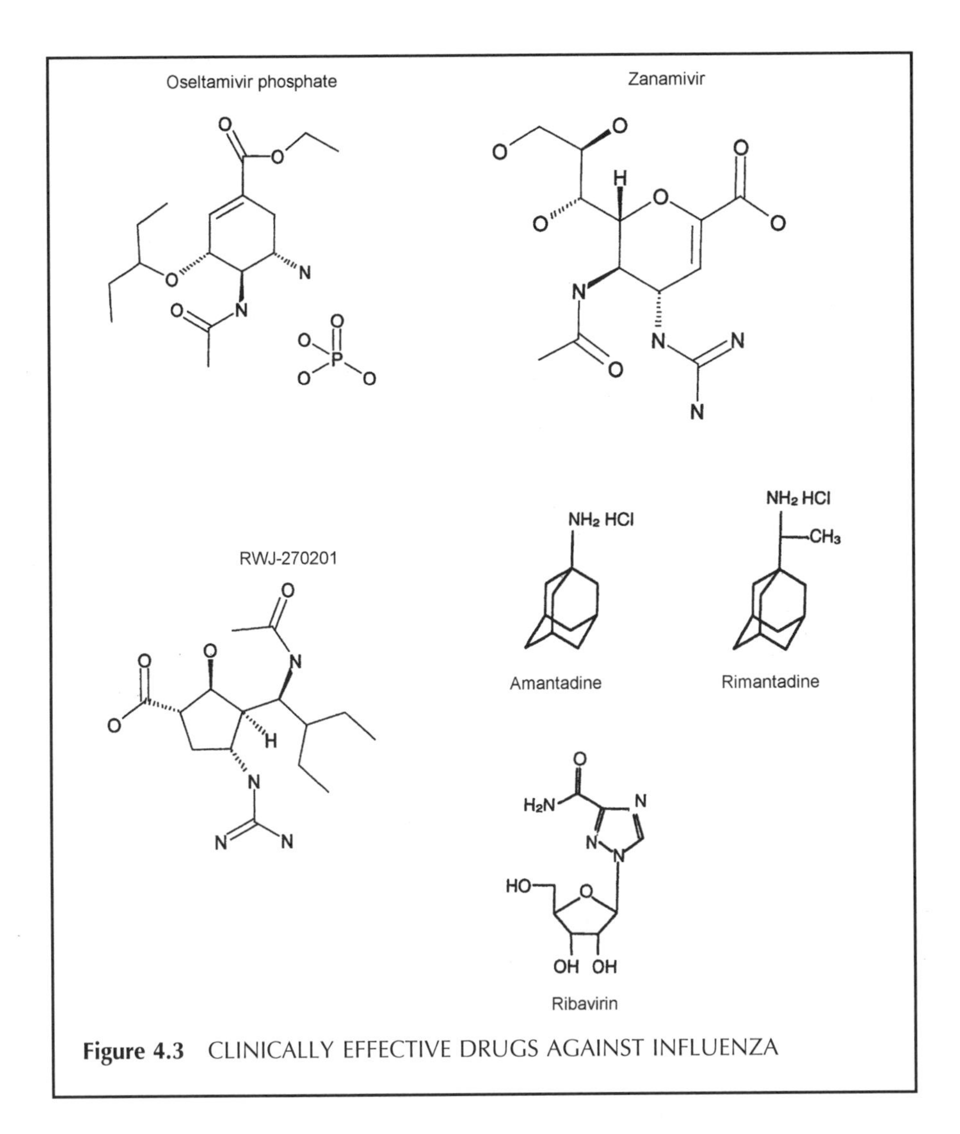

Figure 4.3 CLINICALLY EFFECTIVE DRUGS AGAINST INFLUENZA

these drugs, including non-nucleoside inhibitors of reverse transcriptase and protease inhibitors, are listed in Table 4.2. The inhibitors of HIV protease are used in combination with AZT and, perhaps, ddI or ddC as a drug combination. Also a prodrug of aciclovir called valaciclovir has been developed. This is the L-valine ester of aciclovir and after absorption undergoes almost complete hydrolysis to aciclovir and the essential amino acid L-valine. Valaciclovir itself has negligible pharmacological activity and all products of its metabolism except for aciclovir are inert or well characterized. Of course the most important product of its metabolism is aciclovir itself. The crucial point is that the oral prodrug leads to 3–5 times enhanced bioavailability of the active

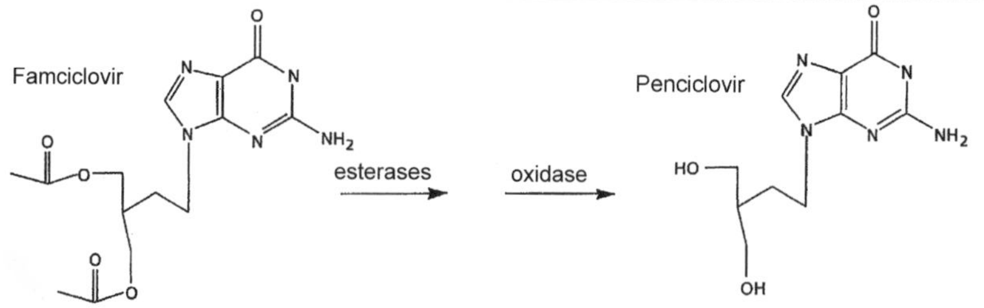

Famciclovir

Penciclovir

esterases

oxidase

Figure 4.4 FAMCICLOVIR AS A PRODRUG OF PENCICLOVIR

aciclovir, enabling less frequent doses and higher plasma levels. Similar in concept is the clinical application of the prodrug famciclovir which is converted enzymatically in the patient to the active anti-herpes nucleoside analogue drug penciclovir (Figure 4.4). An expanding focus is with hepatitis B virus where an estimated 5% of the world are chronic carriers of the virus and combination chemotherapy with lamivudine, adefovir and famciclovir is being tested experimentally.

ACICLOVIR AND OTHER ANTI-HERPES DRUGS

Nucleoside analogue predecessors of aciclovir such as idoxuridine (IDU) and trifluorothymidine (TFT) were useful for treating superficial herpes infections (including those of the eye), whilst adenine arabinoside (ara A) had an important and pioneering role in the treatment of herpes encephalitis and serious paediatric infections. However, the advent of aciclovir transformed the often difficult clinical management of herpetic infections and these earlier discovered drugs, with the possible exception of TFT, are no longer used. Aciclovir is used as a prophylactic, before surgery for example, to prevent recurrent herpes type I infections in bone marrow and heart transplant patients, and therapeutically to prevent spread of mucocutaneous infections in already infected and immune compromised persons. It has also been shown to be very effective in saving lives when used to treat herpes encephalitis. Aciclovir is also used against recurrent HSV infections, particularly those of the genital tract, and to a lesser extent in various forms of VZV infections. Actually there is an urgent need for highly active drugs against VZV and this infection will come even more to the public's notice as the population ages. More effective against VZV is the prodrug of aciclovir, valaciclovir which is better absorbed orally and is rapidly converted to aciclovir *in vivo*.

A molecular relative ganciclovir, has antiviral activity against cytomegalovirus (CMV) and is used to treat life-threatening CMV infections after bone marrow transplants, or CMV pneumonia or retinitis in AIDS patients. However, unlike aciclovir the drug induces rather severe neutropenia and thrombocytopenia. Although CMV does not have a viral TK enzyme, another viral protein phosphorylates ganciclovir to the monophosphate and thereafter a cellular enzyme phosphorylates the molecule to ganciclovir triphosphate, which inhibits CMV DNA polymerase.

Foscarnet, which is not a nucleoside analogue, can be used to treat intractable cases of CMV but the drug is administered in hospitals under close clinical care.

Aciclovir possesses an excellent combination of pharmacological and antiviral properties which helps to explain its unique and highly specific anti-herpesvirus specificity. Firstly, the compound is only phosphorylated in herpes-infected cells, since the herpes thymidine kinase enzyme is less 'precise' than the corresponding cellular enzyme and will accept fraudulent substrates, such as aciclovir. This means that any potential drug-induced toxicity is immediately avoided or at least confined to a virus infected cell. Once phosphorylated to the triphosphate by cellular enzymes the latter molecule inhibits, again specifically, the function of the herpesvirus DNA polymerase and has no effect on cellular DNA polymerases. It is both an inhibitor and a substrate of viral DNA polymerase competing with GTP and being incorporated into viral DNA. It has little or no effect against cellular DNA polymerase. New herpes DNA elongation is aborted very successfully because aciclovir lacks the 3′ hydroxyl group on the sugar ring required for phosphate–sugar linkage and hence addition of new bases and hence DNA chain elongation.

However, a latently infected cell cannot be cured and thus aciclovir does not eradicate herpesvirus from an infected individual, but it can be used to prevent clinical recurrences. The compound has proved to be remarkably safe in clinical practice and some patients have taken the drug orally (daily) for several years. In fact in several countries in the European Community, aciclovir can be purchased as an 'over-the-counter' drug for self-prescription for cutaneous HSV infection, as with cold sores around the mouth.

Another nucleoside analogue closely related to aciclovir and called penciclovir is widely used in clinical practice, with an advantage that fewer daily doses are required. As with the prodrug valaciclovir, a prodrug of the new molecule has been introduced called famciclovir (Figure 4.4). This prodrug may also have clinical usefulness against hepatitis B virus.

AMANTADINE, AN M2 CHANNEL BLOCKER OF INFLUENZA A VIRUS

This cyclic primary amine was discovered by chance as an influenza virus inhibitor in the 1960s. It had in fact been synthesized as a potential explosive and not as a biological. Numerous clinical trials have shown that prophylactic administration as a tablet given twice a day will prevent influenza A (H3N2, H1N1 or H2N2) clinical infection in 70–80% of individuals. Unfortunately it has no inhibitory effect against influenza B viruses which are important viruses causing mortality about every fourth year. Nervousness is the main side-effect noted in about 8% of persons receiving 200 mg amantadine per day but dosage can be halved to avoid these problems and still maintain antiviral effects. A very similar molecule but with an extra methyl group attached (rimantadine) has equivalent clinical activity but causes rather fewer side-effects.

Studies of the therapeutic use of amantadine in prisons, schools and universities throughout the world showed, perhaps surprisingly, that if the drug was given after infection but within 24–48 hours of the onset of symptoms these resolved more quickly and the number of days of incapacity was reduced. Thus therapeutic use of a drug against a respiratory virus is possible, and this opportunity, unexpected at the time, has been exploited by the newer antineuraminidase drugs.

Recommendations from a WHO expert group are that the anti-influenza compounds should be used prophylactically where epidemiological investigations show the presence of influenza A in the community. Prophylactic use should continue daily for up to 4–5 weeks until the epidemic has passed. Chemoprophylaxis is recommended for 'special-risk' groups, such as over-65s, some diabetics, and persons with chronic heart or chest diseases who have either not been immunized or who wish to receive additional protection to that of immunization, or those being potentially exposed to virus infection before vaccination has become effective (2–3 weeks). These members of the community are at much higher risk of serious complications and death following influenza than others. Clinicians now appreciate that amantadine dosage must be carefully adjusted for elderly and frail individuals and particularly those with kidney disease or urinary retention, in whom the drug could accumulate. A reduced dose of 100 mg or lower daily is recommended.

The mode of action of the drug is still the subject of research and the principal viral target is the viral M2 protein. The M2 protein forms a transmembrane ion channel in the virus and amantadine can block this channel much as a gate can prevent access to the entrance of a building. When the M2 ion channel is functioning correctly it allows the passage of hydrogen ions to the centre of the virus and hence acidification when the virus is being uncoated. Under the influence of the resulting low pH inside the virus the M protein dissociates from the viral RNA which is then free to infect the nucleus of the cell. All these vital early stages of viral uncoating are blocked by amantadine when it sits in the ion channel and blocks off access for protons. It is no surprise that drug resistance to amantadine occurs when mutations in the M2 gene and subsequent amino acid substitutions in the M2 protein prevent the binding of the molecule and hence stop its effect as a gate.

RELENZA AND TAMIFLU, INHIBITORS OF INFLUENZA A AND B NEURAMINIDASE

The new anti-influenza drugs have a significant advantage compared with amantadine of inhibiting both influenza A and B viruses. Relenza is rather poorly absorbed after oral dosing and has to be administered by dry powder inhaler, but Tamiflu may be administered by mouth. These two new drugs could revolutionize the way in which influenza is managed in clinical practice and thereby speed patients' access to effective chemotherapy.

Both drugs bind to a group of 11 or so amino acids in the active site of the NA enzyme. These amino acids are constant in all current influenza A and B viruses and so the drugs inhibit all these viruses. Even previous pandemic viruses such as the Great Pandemic of 1918 have a near identical active site and are inhibited. Drug-resistant mutants have been described but to date appear less pathogenic and less infectious than the wild-type virus and thereby would not be expected to spread in the community. Although research is continuing with anti-common cold drugs none to date has shown strong enough clinical effects to warrant extensive use and virtually no drugs exist for the remaining important respiratory viruses, namely adenoviruses, parainfluenzaviruses or coronaviruses.

A nucleoside analogue, ribavirin, originally researched as an anti influenza drug is licensed to treat children with severe respiratory distress after infection with respiratory syncytial virus, but the necessity of an aerosol apparatus not unnaturally has restricted the usefulness of the drug.

ZIDOVUDINE AND DIDEOXYNUCLEOSIDE ANALOGUES AS ANTI-RETROVIRUS DRUGS

Within 2 years of the isolation of the AIDS virus a series of promising antiviral compounds had been discovered and zidovudine had been shown to be effective in prolonging the life of AIDS patients. The drug is rapidly absorbed after oral administration with a short 1 hour half-life and so the drug is given two or three times daily. The first double-blind placebo-controlled trial of zidovudine in AIDS patients had to be stopped and the code broken when it was found that virus induced lethality in the zidovudine-treated group was only one compared with 19 in the control group of approximately 194 patients. In addition, fewer opportunistic infections developed in the zidovudine-treated group, and a reduction in the level of circulating viral p24 core antigen indicated a specific antiviral effect of the compound. Zidovudine is now considered to be useful for prolonging the life of these patients for up to 1 year, but its use is enhanced by drug combinations. The original severe problems of toxic effects of the drug on bone marrow cells necessitating transfusion in approximately one-third of the patients have been overcome both by dosage reduction and use earlier in the disease when the patient is still fit. However, headaches, nausea and insomnia are not uncommon side-effects of the drug and the patients have to be carefully monitored.

Zidovudine and the other dideoxynucleoside analogues are potent inhibitors of viral reverse transcriptase and hence DNA synthesis and like aciclovir are able to prevent chain elongation by, in the case of zidovudine, the 3′ positioning of the azido group. Therefore once the drug is incorporated into the newly growing viral DNA strand the essential phosphodiester–diester linkage enabling the next nucleotide to be added to the growing DNA chain is blocked. The dideoxynucleoside monophosphate is phosphorylated to the triphosphate, by cellular enzymes, and the triphosphate differentially inhibits the viral reverse transcriptase enzyme and has lesser effects on the cellular DNA polymerase. Unfortunately the phosphorylation to the active triphosphate occurs not only in virus-infected cells but also in normal cells, and this explains the side-effects of the drug. In contrast, as we have noted above, aciclovir triphosphate is only present in herpes-infected cells.

A number of dideoxynucleoside analogues (ddI, ddC, 3TC, D4T) have been clinically evaluated and although they all exert antiviral effects nevertheless they also have side-effects, each specific to the compound in question. Also effective are non-nucleoside analogue inhibitors of viral reverse transcriptase and particularly inhibitors of other viral target enzymes such as protease. Work has also commenced to find drugs to interact with important viral proteins such as integrase and regulatory proteins which control the speed of viral replication, such as *Rev* and more recently fusion functions of the gp 41 portion of the spike protein.

Non nucleoside and protease inhibitors for HIV

Most importantly, combination chemotherapy with other dideoxynucleoside inhibitors (ddC, ddI, 3TC, D4T), or alternatively with non-nucleoside RT inhibitors such as nevirapine or viral protease inhibitors (saquinavir) is now widely used in AIDS patients to reduce the chance of drug resistance (HAART). Currently there is some conflicting evidence from clinical trials whether zidovudine and drug combination therapy should be used early in the disease in asymptomatic patients or whether it should be reserved until the patients begin to show early clinical signs of immunosuppression.

THE MOST RECENT PAST AND THE FUTURE

Arguably the most important chemotherapeutic advances in the last years have been in the treatment of AIDS patients with drug combinations (HAART). Combination chemotherapy has now become the accepted clinical approach with HIV and will be rapidly extended to include new drugs and may be used in cases of chronic hepatitis B and C infection and influenza. Undoubtedly a clinical breakthrough has been the development of the two prodrugs, valaciclovir and famciclovir, to treat herpes infections. Fortunately drug resistance is not likely to be a major problem with herpesviruses. Two important antineuraminidase drugs against influenza have been licensed and a further two compounds are in development. These new possibilities in the clinical management of influenza have led to a renewed interest in the first antiviral, amantadine.

There also remain some very important target viruses against which few, if any, antivirals exist. Thus, the eight hepatitis viruses all cause serious disease whilst we now appreciate that the neglected papillomaviruses could be an important cause of skin cancer. New herpesviruses continue to be discovered, whilst only HHV1 and 2 are significantly inhibited by the existing drugs. Threats of bioterroism will lead to evaluation of drugs against smallpox such as cidofovir and against viruses which cause haemorrhagic fevers such as Lassa and Ebola. Influenza has recently been added to the list of potential bioterrorist viruses.

Scientifically, major drug discoveries are most likely to emerge from a judicious mixture of biology and X-ray crystallography whereby existing compounds are closely analysed for their binding or target interactions at the molecular level. Already this approach has led to the refinement of the two drugs binding to influenza A and B neuraminidase, protease inhibitor of HIV and a drug against the common cold virus. Virologists are anticipating a new influenza pandemic and therefore antivirals with a broad antiviral spectrum would be comforting to have. Amantadine itself does inhibit most, if not all, of the influenza A subtypes known to exist in humans, birds, pigs and horses and so might be expected to inhibit even a new pandemic influenza A virus. The new neuroaminidase inhibitors would also act against pandemic influenza A virus.

Four decades ago the discovery of what we now recognize as important cytokines, α and β interferon, led to over-hasty conclusions that viruses could be conquered by these broad spectrum molecules. Interferons continue to be carefully investigated as potential inhibitors of hepatitis B and C viruses, often in conjunction with specific antivirals such as ribavirin and famciclovir, and if successful in this role may yet fulfil their early promise.

There will be occasional major setbacks, but antiviral chemotherapy like other clinical-based sciences will continue to benefit from exciting scientific discoveries, new clinical advances and further knowledge of the pathology of disease.

IS THERE A BETTER WAY?

5. VIRUS VACCINES

Lat. vacca = cow; *vaccinia* = cowpox.

L. R. Haaheim and J. R. Pattison

THE GLORIOUS PAST... AND THE NEW CHALLENGES

The global eradication of smallpox stands as a landmark in the history of immunization. An intense combined international effort, The Smallpox Eradication Programme of 1967 organized through the World Health Organization (WHO), led to the complete elimination in 1977 of one of mankind's great scourges. A long time had passed since Jenner in 1796 successfully inoculated a farmer's boy with pustule material from a dairymaid suffering from cowpox (see illustration to Chapter 41). Additional achievements during the 20th century have been the introduction of many new virus vaccines, e.g. polio, measles, rubella, mumps, rabies, yellow fever, influenza, varicella and hepatitis A and B. Vaccines against e.g. adenovirus, cytomegalovirus, herpes simplex virus and rotavirus are currently being developed or are under clinical trials.

Great optimism followed the successful smallpox programme, but not all infectious diseases of viral aetiology may be so amenable to control. In particular, a vaccine against HIV may be more difficult to develop than at first anticipated. Nevertheless, there is a range of virus diseases within reach of being controlled by vaccination at least in developed countries. Poliomyelitis is for all practical purposes eliminated from many countries through extensive use of either live (attenuated) or killed (inactivated) vaccines among children. In some developed countries measles is now close to eradication or is an extremely rare disease, and through continued mass vaccination mumps and rubella may also become infections of the past. On the other hand, influenza is difficult to control by vaccination since the virus changes its antigenic properties and previously acquired immunity becomes obsolete. It may be even more difficult to control such illnesses as the common cold because of the multiplicity of viruses involved. For example, there are over 100 serotypes of rhinoviruses and these are difficult to replicate in the laboratory.

It is notable that those virus infections which have been well controlled by immunization are systemic infections in which a viraemia is an essential component of the pathogenesis of the disease. Much greater difficulty has been experienced in developing effective vaccines against superficial mucosal infections of the respiratory and gastrointestinal tracts and against those diseases in which the virus remains largely cell-associated. This latter problem may well apply to the development of HIV vaccines.

ACTIVE IMMUNIZATION

Vaccines are preparations, administered either orally or parenterally, which stimulate a protective specific immune response in the recipient without themselves causing disease. With some vaccines (e.g. hepatitis B) it is necessary to add an adjuvant which non-specifically potentiates the immune response. Adjuvants, usually aluminium salts, can also delay the release of vaccine material from the injection site. In principle there are two main types of viral vaccines available, namely the **attenuated** live virus vaccine and the non-replicating (**'killed'**) variety. In the first instance virus is manipulated *in vitro* to be of low virulence but able to replicate in the vaccinee and stimulate the desired protective immune response without causing disease. In the case of non-replicating vaccines the virus may be replicated in an appropriate laboratory system, e.g. cell culture or embryonated eggs, purified and finally inactivated ('killed') by chemical means (usually formalin). A further refinement of the inactivated vaccine is to split the virions by means of detergents, or to isolate the desired viral subunits to make up the final vaccine. Hepatitis B vaccine is non-replicating and obtained by the expression in yeast of cloned viral genes.

To date, emphasis has been placed on live virus vaccines except where laboratory replication has not been possible (hepatitis B) or satisfactory attenuation has not been achieved (rabies). Less antigenic mass is required in a live virus vaccine since there is replication and build-up of immunogen in the vaccinee. As a consequence inactivated vaccines are usually more expensive. Live virus vaccines induce long-lasting immunity after a single dose (live poliovirus vaccine requires three doses to ensure protection against all three types); killed vaccines often require multiple doses. Live virus vaccines also have their drawbacks. Through the manufacturing process adventitious agents may be incorporated. Attenuated vaccine strains may revert to virulence on passage through the human host. Neither of these problems have proved significant in practice. Finally, attenuation is judged in immunocompetent individuals, therefore live vaccines are often contraindicated in immuno-compromised individuals and also during pregnancy.

It is generally believed that complete protection against infection is extremely difficult to obtain by vaccination, or at least not for any long period of time post-vaccination. The initial hope for a HIV vaccine was precisely that: to confer complete protection (sterilizing immunity) and thus arrest the virus at the site of entry. This may indeed be a formidable task.

However, there is abundant evidence that many vaccines elicit an excellent protection against disease. Thus, the actual infection may take place, and possibly generate a beneficial booster immune response. The infection will nevertheless be quickly aborted due to the immunological recall of the vaccine-induced memory, and/or because of some level of pre-existing active immunity not necessarily located at the site of viral entry. A good example of such a case is the use of inactivated poliovaccines that in some Scandinavian countries has managed to eliminate poliomyelitis, even if the post-vaccination intestinal immunity has been low or absent.

PASSIVE IMMUNIZATION

So far we have spoken only of active immunization, but passive immunization also has a place in preventing virus infections. Passive immunization is the transfer to one individual of antibodies formed in another. The advantage is that it is rapid in onset (effective within a day), but it has the disadvantage of being short-lived (lasting only 2–6 months) since the injected foreign antibody decays with a half-life of 21 days. Human immunoglobulin preparations are made of donated blood through a series of fractionation steps. If they are made from unselected blood donors the preparation will be rich in specific antibodies that are common in the population. Hyperimmune globulins are made from a pool of units of blood selected because they have a high titre of a particular antibody. Such preparations are available for the prevention of hepatitis B, rabies and chickenpox.

In certain situations (rabies, neonatally acquired hepatitis B) it is possible to combine active and passive immunization and take advantage of the best features of each, the immunoglobulin giving immediate protection until lasting active immunity from vaccination develops.

ADMINISTRATION OF VACCINES

The route of administration depends on the vaccine. Live polio vaccine is always given by mouth. Injectable vaccines are usually administered intramuscularly or deep subcutaneously, although an equally effective but smaller dose can sometimes be given intradermally and this may be relevant with expensive vaccines (e.g. rabies).

Since vaccination usually requires attendance at a clinic it is often worthwhile giving more than one vaccine at a visit. Individuals can respond to multiple antigens administered simultaneously, although between 4 and 14 days after one vaccine individuals may respond poorly to another. Therefore vaccines should either be given simultaneously, preferably at different sites, or after an interval of at least 3 weeks. If multiple doses are required the optimum interval will depend on which type of vaccine (live or killed) is used. It is 1 month between doses of polio vaccine and 1 and 4 months, respectively, between the first and second and the second and third doses of hepatitis B vaccine. However, intervals may vary and the reader is advised to consult national regulations.

SIDE-EFFECTS AND COMPLICATIONS

It is important to remember that trivial side-effects are quite common with viral vaccines. These are most often local pain, redness and induration at the injection site. Less frequent are systemic effects such as fever, malaise, headache, arthralgia and nausea. With some live virus vaccines such as measles and rubella mild symptoms and signs resembling the natural disease may occur.

Serious complications are extremely rare, but they do occur; for example, paralytic poliomyelitis following the use of live attenuated polio vaccine, and encephalitis following yellow fever vaccination in infants. Note that it is important to adhere to the manufacturer's/health authority's list of contra-indications cited for each vaccine in question so that the risk of allergic reactions or other preventable complications can be minimized.

Immunosuppression is a contraindication for use of live vaccines. However, different countries are adapting different strategies for the immunization of HIV-positive children depending upon the relative risk of side-effects of the vaccine and danger from the disease being immunized against. As a rule, it appears that asymptomatic HIV-infected individuals can be offered virus vaccines.

EPIDEMIOLOGICAL CONSEQUENCES OF MASS VACCINATION

If an individual benefits from being immunized it is often assumed that the overall effect on society at large will be equally beneficial. This may not, however, always be the case. The reason for this is that we are changing aspects of herd immunity and it may not be possible to precisely anticipate the shift of susceptible cohorts. Deferring childhood diseases to older age groups may consequently increase the frequency of complications if the vaccine coverage is too low (e.g. measles).

A special case is the use of rubella vaccine: if only a low but still significant fraction of children and teenagers is immunized, there is a possibility, through the consequent reduced spread of wild-type virus in the community, of accumulating more susceptible women at childbearing age, and thus possibly increasing the incidence of congenital rubella syndrome.

Another special case is the selective use of influenza vaccines among risk groups aimed at reducing morbidity and mortality among these individuals rather than attempting to eliminate the virus from society. Indeed, eradicating influenza through mass vaccination is for many reasons believed to be impossible.

To avoid epidemiological backlash it is for many vaccines often necessary to achieve very high (>90%) rates of immunization. This can only be done with a high level of acceptance by the community, combined with great efforts on the part of the health care system and its officers. These factors are often present when many cases of a serious disease occur, but interest tends to wane once the disease becomes rare. The public at large will then inevitably focus on the possible side-effects and complications of the vaccine preparation itself, rather than the morbidity and mortality that the disease would cause in an unprotected society. With time, the public will forget the effect that these diseases had on health and welfare, and there may be an increased tendency for doctors and health workers to grant exemption, thus reducing vaccine coverage. Some concern has been raised claiming that the discontinuation of smallpox vaccination leaves the world susceptible to reintroduction of poxviruses from unrecognised animal reservoirs. And similarly, an immunologically virgin global population could invite terrorist threats or the actual release of variola virus procured from illegal stocks of virus.

Another case is the eventual elimination of measles virus and the subsequent end to measles vaccination programmes. The consequential waning and disappearance of herd immunity may allow canine distemper(-like) viruses to be introduced from animal reservoirs.

In order to maintain a high immunization rate over the years it is of paramount importance to continuously inform the public and the health professions alike of both the individual and community benefits of well designed and implemented mass programmes of immunization. Failure to do so will inevitably lead to a reduced coverage, the accumulation of new susceptibles thus paving the way for new outbreaks.

THE FUTURE

Considering the glorious past it is tempting to postulate an equally bright future for vaccines and vaccine programmes, especially since modern science and technology seem to advance at such high pace. But even today's technology has not been fully exploited. It is assumed that within a few years both polio and measles may become extinct as human diseases. This ambitious global effort is directed by the WHO.

Modern science has provided a better understanding of the fine structures of virus particles, thus allowing the identification and subsequent large-scale production of the most useful parts of the infectious agent for vaccine use. This methodology has now been put to good practical use for the preparation of hepatitis B vaccine using cloned DNA expressed in yeast cells to produce HBsAg. This was a giant leap from the earlier method of purifying HBsAg from the blood of hepatitis carriers.

Other strategies being investigated are the use of synthetic oligopeptides covering important epitopes on the virus or peptides masquerading as epitopes ('mimotopes'). However, since the immunogenicity of many such preparations may fall short of the ideal, these strategies would require new and more potent adjuvants than are currently in use, as well as safe carrier molecules. Other approaches include investigating ways of stimulating mucosal immunity by targeting vaccine material to mucosal epithelium, or using biodegradable microspheres that can release vaccine material in a controlled way. Possibly the most promising and certainly daring strategy is to inject 'naked' DNA, in the form of a non-replicating plasmid coding for selected viral proteins, into e.g. muscle tissue of the vaccinee and let the injected DNA code for proteins in an authentic form to be properly presented on MHC molecules. This approach will in many ways mimic the process of a viral infection and thus stimulate both humoral and cell-mediated immune response. For manufacturers these modern methods avoid large-scale production of potentially harmful infectious agents. However, clinical trials so far have not lived up to its promises. Another unexpected strategy is the large-scale production of viral proteins in plant cells ('edible vaccines').

Many other questions need to be addressed:

- Which are the most cost-effective ways of vaccine production and delivery?
- How can we improve vaccine stability and storage, especially in the tropics?
- Can we reduce the number of visits to health clinics by developing more multivalent vaccines?
- Can reliable methods of time-controlled slow release of vaccine material be developed, ensuring both priming and booster doses at a single visit to the clinic?
- More effective adjuvants and vaccine vehicles should be developed. Especially important are designed molecules that can ensure uptake by epithelial cells at the mucosal linings.
- How can a more potent mucosal response be stimulated?
- How can we ensure better long-term immunological memory?
- How can we more effectively stimulate CTL response for inactivated vaccines?
- Using 'naked' DNA as vaccine material, can we ensure that no integration of the injected DNA occurs and that no adverse immunological reactions take place through prolonged antigenic stimulation?

Both scientific and technical progress is needed, but equally important is a well-structured community health service and public motivation to take advantage of the many excellent vaccines already available. See Table 5.1.

Table 5.1 SOME CURRENTLY AVAILABLE VIRUS VACCINES

Vaccine	Nature	Route**	Timing
Polio	Live attenuated	Oral	Infancy/childhood
	Inactivated	s/c or i/m	Similar (or when live vaccine is contra-indicated)
Measles*	Live attenuated	s/c or i/m	Infancy/childhood
Mumps*	Live attenuated	s/c or i/m	Infancy/childhood
Rubella*	Live attenuated	s/c or i/m	Infancy/childhood/adolescent girls, susceptible women post-partum
Influenza	Inactivated	s/c or i/m	The elderly and those with certain chronic diseases
Hepatitis A	Inactivated	i/m	Travellers, occupational exposure
Hepatitis B	Inactivated	i/m	High-risk groups, occupational exposure
Rabies	Inactivated	s/c, i/m or i/d	Occupational exposure, post-exposure treatment
Yellow fever	Live attenuated	s/c	Travellers
Japanese encephalitis	Inactivated	s/c	Travellers
Tick-borne encephalitis	Inactivated	i/m	Travellers
Varicella	Live attenuated	s/c	Immunocompromised
Vaccinia	Live	i/d	Laboratory workers handling smallpox virus

*Available in combination as MMR vaccine.
**s/c, i/m, i/d stand for subcutaneous, intramuscular and intradermal, respectively.

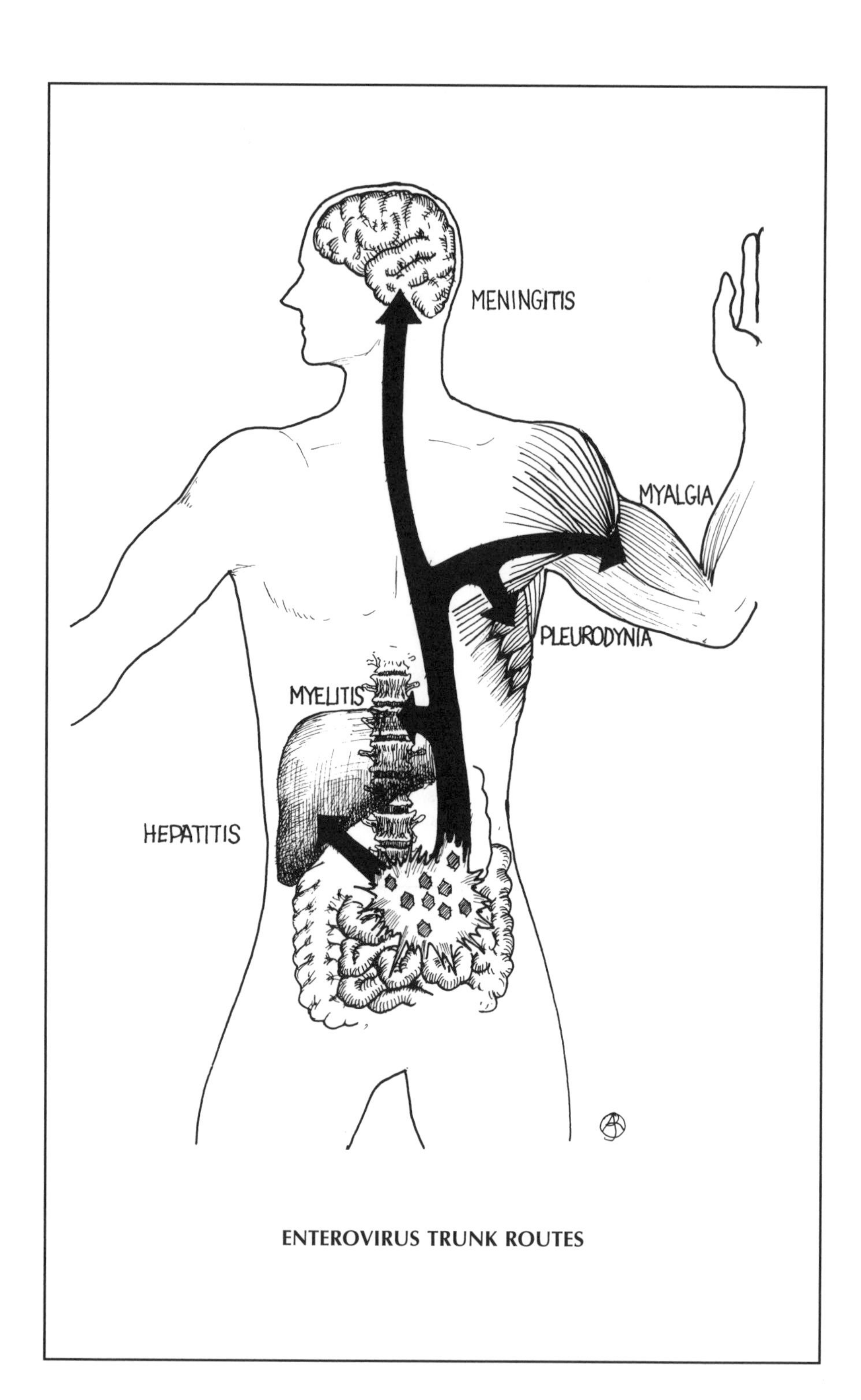

ENTEROVIRUS TRUNK ROUTES

6. ENTEROVIRUSES: POLIOVIRUSES, COXSACKIEVIRUSES, ECHOVIRUSES AND NEWER ENTEROVIRUSES

Gr. *enteron* = small intestine, the main replication site for most enteroviruses.

A.-L. Bruu

The *Enterovirus* genus of the picornavirus family is a large group of viruses associated with a spectrum of diseases ranging from paralytic poliomyelitis to mild, non-specific febrile illness and rarely associated with disease of the gastrointestinal tract. They are worldwide in distribution but more than 90% of infections with enteroviruses are subclinical.

The *Enterovirus* genus comprises several subgroups of which the following may cause disease in humans:

- **Polioviruses** (types 1–3). Gr. *polios* = gray, *myelos* = marrow.
- **Coxsackieviruses**, Group A (types 1–22, 24) and Group B (types 1–6). Coxsackie is the village in the USA where the patients from whom these viruses were first isolated lived.
- **Echoviruses** (types 1–9, 11–27). Enteric cytopathogenic human orphan viruses, originally considered not to be associated ('orphan') with human disease.
- **Newer enteroviruses** (types 29–34, 68–72). Human enterovirus 72 is hepatitis A virus, see Chapter 24.

The enteroviruses have a diameter of 24–30 nm, an icosahedral structure and consist of 60 subunits, each containing one set of the structural proteins VP1–4. The single-stranded RNA has positive sense (mRNA function). The complete nucleotide sequence has been determined for the polioviruses and some other enterovirus types. Some enteroviruses may cross-react to a certain degree, mainly due to determinants on VP1.

Clinical syndromes frequently associated with specific types of enteroviruses include the following:

- Paralytic disease: polioviruses.
- Herpangina: coxsackie A viruses.
- Hand, foot and mouth disease: coxsackie A virus (A16).
- Epidemic myalgia/pleurodynia: coxsackie B viruses.
- Generalized disease in the newborn: coxsackie B viruses.
- Myocarditis/pericarditis: coxsackie B viruses.
- Conjunctivitis: enterovirus 70.
- Fever and rash: echoviruses especially.
- Meningitis: many enteroviruses.

1400BC EGYPTIAN STELE SHOWING PRIEST WITH 'HORSE-FOOT'. POLIOMYELITIS? (Courtesy of Ny Carlsberg Glyptotek, Copenhagen)

7. POLIOVIRUSES

Infantile paralysis; acute anterior poliomyelitis; Ger. *Kinderlähmung*.

A.-L. Bruu

Poliomyelitis is an acute infectious disease with or without signs of CNS involvement.

TRANSMISSION/INCUBATION PERIOD/CLINICAL FEATURES

The infection is spread by the faecal–oral route. The incubation period is usually 1–2 weeks. The patient can infect susceptible persons from some days before illness and for one to several weeks after the illness. Children are infectious for a longer period than adults.

SYMPTOMS AND SIGNS

Systemic:	Fever, Headache, Myalgia, Nausea, Vomiting
Local:	Signs of Meningitis, Pareses

About 95% of infections run a subclinical course. Patients suffering from aseptic meningitis will recover in 1–2 weeks. Paralysis often results in persistent lameness.

COMPLICATIONS

Respiratory failure, obstruction of airways, involvement of the autonomic nervous system.

THERAPY AND PROPHYLAXIS

No specific therapy, immunoglobulin is of no practical value. Vaccine, either attenuated or inactivated, gives more than 90% protection.

LABORATORY DIAGNOSIS

Demonstration of poliovirus in throat swab or in faecal sample collected in the acute phase of the disease, viral RNA in faecal sample, and poliovirus IgM antibodies or IgG antibody rise in paired sera.

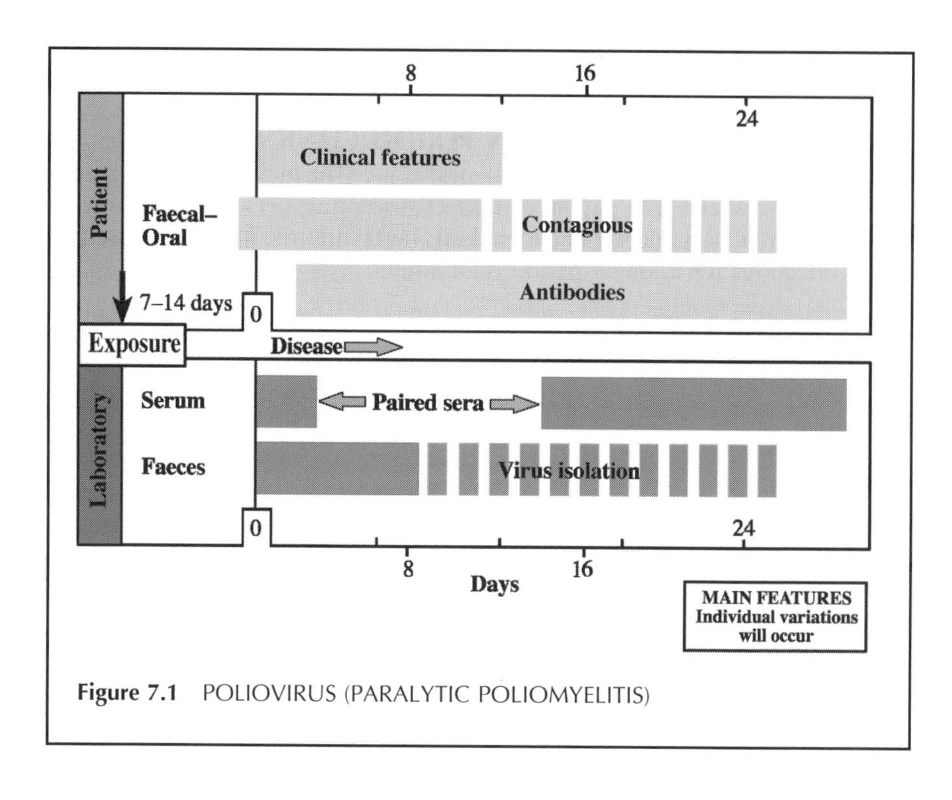

Figure 7.1 POLIOVIRUS (PARALYTIC POLIOMYELITIS)

CLINICAL FEATURES

SYMPTOMS AND SIGNS

The incubation time of poliomyelitis is usually 7–14 days, but may vary from 4 to more than 30 days. The disease typically starts with a prodromal phase of a few days' duration. The patient has fever and complains of myalgia. Constipation is a common feature. This phase ('minor disease') is usually followed by an interval of a few days when temperature becomes normal and the patient seems to recover. The temperature then increases again with the development of paralysis and frequently also aseptic meningitis. Such a biphasic course is especially common in children. The second phase is initially characterized by hyperirritability and increased tendon reflexes. This may last from several hours to a few days, leading to the paralytic stage with loss of tendon reflexes. The paralysis is flaccid and most frequently affects the extremities, but any voluntary muscle (group) may be involved. The development of paralysis may take some hours or a few days. During the initial phase of the paralytic stage the patient may also exhibit sensory disturbances. Strenuous exercise, injections (vaccination), operations (tonsillectomy) and possibly also pregnancy may increase the incidence, severity and site of paralytic disease. **Bulbar poliomyelitis** may occur alone (in about 10% of all patients with paralysis) or as a mixed bulbospinal form. This localization may lead to involvement of cranial nerves with paralysis of pharyngeal muscles and dysphagia, and of respiratory muscles followed by dyspnoea. Bulbar involvement is often accompanied by lesions of the respiratory and circulatory centres leading to respiratory failure, fall in blood pressure and circulatory shock. The lethality of this condition varies between 20 and 60%. The CSF often shows normal values, but an increase in cell count up to a few hundred/ml is sometimes seen. Polymorphonuclear cells may be prevalent in the very beginning of the disease, but are soon outnumbered by lymphocytes. Slight increase in the protein content may be seen. Immunity after poliovirus infection, whether asymptomatic or paralytic, is type-specific and lifelong.

The most important **differential diagnosis** is polyradiculitis (Guillain–Barré syndrome), where the pareses are ascending and symmetrical combined with a variety of sensory disturbances. The CSF shows a rather high protein content with no or only slight increase in cell count. Other diseases which may mimic poliomyelitis are acute transverse myelitis, tick-borne encephalitis and reduced mobility due to arthritis and osteomyelitis. The diagnosis of poliomyelitis is based upon the development of asymmetrical pareses in the course of some hours to a few days with little or no sensory loss.

CLINICAL COURSE

Fever and general symptoms last for 1–2 weeks. The paralysis reaches a maximum within 2–3 days. More than 50% of cases recover during the subsequent weeks or months. The remaining patients will suffer from residual deficits in one or more muscles. The overall lethality of poliomyelitis has been 5 to 10%, but is substantially reduced by maintaining patients in respirators.

COMPLICATIONS

Encephalitis and myocarditis may occur during the acute stage (see bulbar poliomyelitis). A post-poliomyelitis syndrome is observed in some 25% of survivors of paralytic poliomyelitis. After several decades with no changes in their clinical condition, they develop new weakness, pain and fatigue. This may be due to a denervation of initially reinnervated muscle fibres, but the aetiology is not clear. These patients are not excreting poliovirus and are not contagious.

THE VIRUS

Poliomyelitis is caused by one of the three types of poliovirus (Figure 7.2). The virion is naked and has a diameter of 28 nm. It contains single-stranded RNA of positive polarity (mRNA) within a protein shell (capsid) composed of 60 capsomeres. The capsid is built up of four proteins, VP1–4. Virus replication is initiated by RNA transcription into negative strands to act as templates for new viral RNAs. From the viral RNA a large polyprotein is made, which later is cleaved to generate the capsid proteins VP1–4 and a range of other proteins. Final assembly of new virions takes place in the cytoplasm. There are some minor antigenic cross-reactions between some enteroviruses. Even though the three serotypes of poliovirus share some antigenic properties, in particular between types 1 and 2, they are characterized by marked intertypic differences. The epitopes responsible for inducing neutralizing antibodies are located on the three structural proteins VP1, VP2 and VP3 of the viral capsid, VP1 being the major immunogen. For differentiation between the three types, type-specific antisera prepared by cross-adsorption with heterologous types, or suitable monoclonal antibodies, are used. However, the capsid proteins induce a mainly specific immune response during an infection and after vaccination. All three polioviruses are highly cytopathic to many primary cell cultures and permanent cell lines,

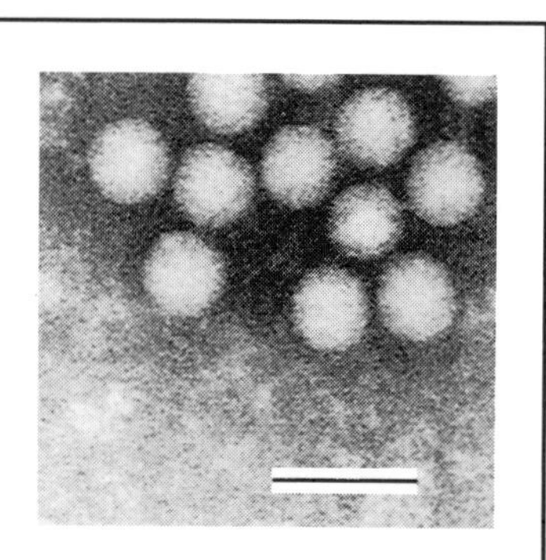

Figure 7.2 POLIOVIRUS. Bar, 50 nm (Electron micrograph courtesy of E. Kjeldsberg)

causing cell death without changes in cell morphology typical of enteroviruses. Polioviruses are stable at pH values between 3 and 9, resistant to lipid solvents and rather slowly inactivated at room temperature. Because of this, the virus may remain infectious for several days in water, milk, food, faeces and sewage.

EPIDEMIOLOGY

Poliomyelitis has probably been with us for centuries. However, it was not until the later part of the nineteenth century that the disease was described as a separate clinical entity. During the first half of the twentieth century several large epidemics of poliomyelitis were observed in Europe and North America. The disease was then most frequent among young children, but in the later part of the period it became more common among older children and adolescents. This was most probably due to improved hygienic conditions reducing the possibilities for faecal–oral spread. In countries with a temperate climate, the disease is mainly seen during summer and autumn months, whereas in tropical and subtropical climates poliomyelitis is prevalent throughout the year and most often occurs in small children. The introduction of polio vaccines in the 1950s has led to more or less complete eradication of poliomyelitis in several countries, especially in Europe and North America. Due to vaccination programmes of small children, most clinical cases are now found among unvaccinated infants, older children and adults. It is therefore important to maintain a high vaccination coverage rate (>90%) to accomplish a sufficiently high degree of herd immunity. Complacency in adhering to vaccination programmes invariably leads to cluster outbreaks of poliomyelitis from imported cases when herd immunity in certain regions or communities comes under a critical limit. In 1988 the 41st World Health Assembly committed the World Health Organization and had set to target the year 2000 as the year of global eradication of poliomyelitis. This goal will probably be reached within the next few years.

THERAPY AND PROPHYLAXIS

There is no specific **treatment** for poliomyelitis. Impairment of respiratory function may necessitate artificial respiration. Physiotherapy as early as possible is important in preventing or reducing lasting sequelae. Although improved sanitation and hygiene help to limit the spread of poliovirus, the only efficient means of preventing paralytic polio is through widespread immunization. Two types of **vaccine** are available against poliomyelitis, inactivated vaccine (IPV, Salk) and live attenuated oral vaccine (OPV, Sabin). Both vaccine formulations contain all three polio types.

OPV is the most widely used vaccine for prevention of poliomyelitis. It is composed of attenuated strains of the three poliovirus types, and is administered orally. At least two or three doses are considered necessary to ensure adequate immunity, in some countries even five to six or more doses are given in the primary course. Revaccination is used to a varying degree. A full primary course induces an antibody response against all three types in more than 90% of vaccinees and gives a high degree of protection against disease. OPV also induces intestinal immunity due to production of secretory IgA antibodies. This is important for inhibition of virus replication in the gut, diminishing the possible spread of virus to susceptible contacts. OPV is almost non-reactogenic, and is very safe. However, in a few cases an attenuated vaccine strain may induce paralytic disease. This occurs in about one case per 1–10 million vaccine doses administered.

IPV was the first vaccine used against poliomyelitis. It contains the three types of poliovirus inactivated by formaldehyde and is administered parenterally. The use of IPV in the late 1950s was followed by a 90% reduction of poliomyelitis cases when it was replaced in many countries by the more easily administered OPV around 1960. Newer IPVs have higher immunogenic potency which has led to a reintroduction of IPV in many developed and developing countries. The primary vaccination course with IPV consists of two or three doses, usually followed by revaccination after intervals of about 5–10 years during childhood and adolescence. Some countries are using a combination of OPV and IPV.

LABORATORY DIAGNOSIS

Recommended methods for laboratory diagnosis of poliovirus infection are:

- **Virus isolation** from faeces and throat washings by inoculation into cell cultures. The presence of poliovirus is shown by degeneration of cultured cells within a few days. The result of conventional typing by neutralization will require another couple of days. Alternatively immunofluorescence using monoclonal antibodies can be used, allowing the distinction between wild-type virus and vaccine strains.
- **Detection of poliovirus RNA** in faecal samples by PCR. This method will also distinguish between wild strains and vaccine strains.
- **Antibody investigations**. The method of choice is μ-chain capture (IgM) ELISA, which is specific for each poliovirus type. Other antibody tests are neutralization and CFT on paired serum samples.

The samples should be collected as early as possible in the course of the disease. Children usually excrete virus for 1–2 weeks, adults for a shorter time. As the excretion may be intermittent during the later phases of the disease, repeated samples should be collected. A negative culture or no poliovirus RNA detected may not exclude infection, especially if the material is taken late in the disease. In such cases, antibody investigations will be useful.

AN ORPHAN VIRUS LOOKING FOR PARENTAL DISEASES

8. COXSACKIEVIRUSES, ECHOVIRUSES AND ENTEROVIRUSES 29–34 AND 68–71

A.-L. Bruu

These enteroviruses may cause febrile diseases, in some cases with signs of infection of CNS, muscle, heart, skin, eye and respiratory tract.

TRANSMISSION/INCUBATION PERIOD/CLINICAL FEATURES

Enteroviruses spread from person to person mainly by the faecal–oral route, and to a lesser degree by the respiratory route. Some types associated with conjunctivitis spread by direct contact. The incubation period is 5–14 (2–25) days. Enterovirus conjunctivitis has an incubation period of 12–24 hours.

SYMPTOMS AND SIGNS

General:	Fever, Headache, Malaise
Neurological:	Meningitis, rarely Encephalitis and Transient Paralysis
Other:	Epidemic Myalgia/Pleurodynia (Bornholm Disease), Myocarditis, Pericarditis, Generalized Disease in the Newborn, Vesicular and Maculopapular Exanthems, Haemorrhagic Conjunctivitis

Usual duration is a few days to about 1 week.

COMPLICATIONS

Occasionally neurological sequelae.

THERAPY AND PROPHYLAXIS

There is no specific therapy, and no immunoglobulin or vaccine against these enteroviruses.

LABORATORY DIAGNOSIS

Virus may be **isolated** from the pharynx early in the disease, from faeces for at least 1 week and in some cases from other sites of infection. Enterovirus nucleic acid may be detected by PCR in faecal sample, throat swab, vesicle fluid, myocardial tissue, pericardial fluid or in cerebrospinal fluid (CSF) for aseptic meningitis. For antibody investigations the μ-capture ELISA is the method of choice.

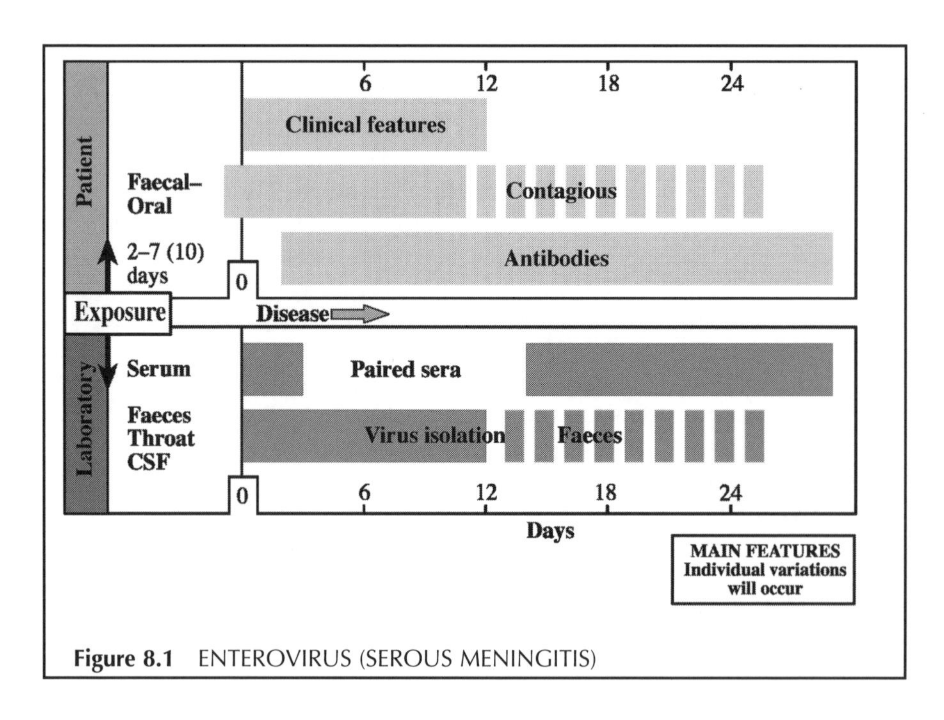

Figure 8.1 ENTEROVIRUS (SEROUS MENINGITIS)

CLINICAL FEATURES

SYMPTOMS AND SIGNS

Like poliovirus the coxsackieviruses and echoviruses multiply primarily in lymphoid tissue in the pharynx and the small intestine. In about 5% of cases virus may spread to other target organs, the main ones being the meninges, the brain and spinal cord, myocardium and pericardium, striated muscles and skin. Infection leads to lasting type-specific immunity. **Fever** of short duration and sometimes a rash or mild upper respiratory symptoms are the most frequent clinical diseases. A few cases progress to one of the following syndromes:

Aseptic meningitis. In typical cases a biphasic course is seen. After an interval of 1–2 days with few or no symptoms, the temperature rises again to 38–39°C, accompanied by headache, neck stiffness and vomiting. A non-specific maculopapular rash, sometimes with petechial elements, may be seen. The CSF is clear with slight or moderate elevation of cell count (up to 500×10^6/litre, mainly lymphocytes) and protein content, but with normal sugar content. The illness may last for 2–10 days, sometimes followed by a convalescent phase of rather long duration. The **prognosis** is good as most patients recover completely. Meningoencephalitis or encephalitis may occur in some cases. In **differential diagnosis**, meningitis caused by other viruses and early or inadequately treated bacterial meningitis which may mimic aseptic meningitis should be considered. Note a petechial rash is also seen in meningococcal disease. Lymphocytes are seen in the CSF (tuberculous, listerial and cryptococcal meningitis), but usually the glucose content is lowered.

Complications are transient paralysis and polio-like disease.

Epidemic myalgia/pleurodynia (Bornholm disease). This is a painful inflammation of the muscles, most pronounced in the intercostal muscles or abdominal muscles, accompanied by pain that may be severe (devil's grip) and resemble ischaemic heart disease or 'acute abdomen'. The pain is often intermittent for periods of 2–10 hours, combined with rise in temperature. The illness lasts for 4–6 days, but relapses in the following weeks are not infrequent. Complete recovery is the rule.

Myocarditis/pericarditis. This is observed in 5% of patients with coxsackie B virus infections. Typical features are fever, chest pain and dyspnoea. Other signs are pericardial rub, heart dilatation and arrhythmias. Heart failure may occur. The illness usually lasts for 1–2 weeks. Relapse may occur during the following weeks and months in 20% of patients. The most important **differential diagnoses** are cardiac ischaemia, infarction and myopericarditis of other aetiology.

Neonatal myocarditis. Some enteroviruses, mostly coxsackie B3 and 4, may cause a severe, often fatal disease in infants characterized by sudden onset, lethargy, tachycardia, dyspnoea and cyanosis. It is a systemic infection as many organs (heart, brain, liver, pancreas) are involved. The virus is transmitted from mother to child just before or at birth.

Herpangina. The illness is seen mainly in children. Some 8–10 vesicles or small ulcers, 1–3 mm in diameter, are seen on the posterior pharyngeal wall. There is pain on swallowing and usually slight fever of a few days' duration. **Differential diagnoses** are herpes simplex, varicella, aphthous stomatitis.
Hand, foot and mouth disease. This occurs most often in children. Moderate fever of 38–39°C may be seen. Vesicles up to 5 mm in diameter are localized on the buccal mucosa and tongue as well as on the hands and feet.
Rashes. Maculopapular rashes ('rubelliform' or non-specific) are seen quite frequently in coxsackie A and echovirus infections, accompanied by pharyngitis and fever. A rash is sometimes seen in the course of meningitis. **Differential diagnoses** are erythema infectiosum, rubella, measles and rashes seen in meningococcal disease.
Acute haemorrhagic conjunctivitis. This eye disease is characterized by pain, swelling of the eyelids and subconjunctival haemorrhages of a few days' duration, usually healing spontaneously in less than a week. It is highly contagious, with an incubation time of 12–24 hours, and spreads by direct contact. Extensive epidemics have been observed in the Far East (caused by coxsackie A type 24) and in Africa, Japan and India (enterovirus type 70). Spread is favoured by poor hygienic conditions as in refugee camps. Associated neurological disease (radiculomyelopathy, cranial nerve involvement) occurs rarely and may lead to residual paralysis.

Coxsackie B has also been associated with idiopathic dilated cardiomyopathy. Some studies have shown evidence for a connection between juvenile diabetes type 1 and coxsackie B virus infection.

THE VIRUS

The enterovirus group (Figure 8.2) is one genus in the *Picornaviridae* family. They are small (28 nm), roughly spherical and contain a single-stranded RNA molecule of positive polarity, which functions as mRNA. The RNA is surrounded by a protein shell (capsid) with icosahedral symmetry. All picornaviruses contain four polypeptides, VP1–4, VP1 being the major immunogen. There is a certain degree of serological cross-reactivity between enterovirus types, especially between types within the same subgroup, due to shared epitopes not exposed at the surface of the virus, as seen when using the complement fixation test. Enteroviruses retain infectivity at pH 3–9 and are resistant to several proteolytic enzymes and lipid solvents. They are stable for days at room temperature. In the laboratory, the coxsackie B and echoviruses will grow in several different cell cultures.

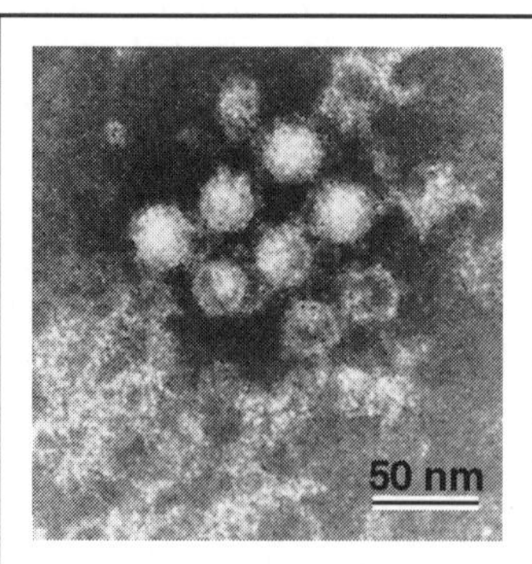

Figure 8.2 ECHOVIRUS WITH ANTIBODY (Electron micrograph courtesy of E. Kjeldsberg)

The coxsackie B viruses will also infect newborn mice. Coxsackie A viruses replicate in mice, but only a few will do so in cell cultures.

EPIDEMIOLOGY

Man is the only natural host for human enteroviruses. The virus replicates in the upper and lower alimentary tract and is excreted from these sites. Enteroviruses spread mainly by the faecal–oral route, and during the acute stage also by the respiratory route. They have a worldwide distribution. In the temperate zones spread takes place in the summer and autumn months, in tropical and subtropical zones throughout the year. Children are infected more frequently than adults, and males somewhat more frequently than females. Poor sanitary conditions will favour spread of these viruses.

THERAPY AND PROPHYLAXIS

There is at present no known specific therapy, nor is there any vaccine against enteroviruses other than the polioviruses. Only symptomatic treatment is available.

LABORATORY DIAGNOSIS

Isolation of virus from stools, rectal swabs, nasopharynx samples, CSF, vesicular fluid and eye secretions has until recently been the most reliable method for laboratory diagnosis of an enterovirus infection. Several types of cell cultures may be used for isolation. Appearance of cytopathic effect (CPE) is observed after a few days, and neutralization tests are used for virus identification. Inoculation of coxsackie viruses into newborn mice will lead to disease and death.

During the last years molecular virological methods such as PCR for the detection of enterovirus nucleic acid (RNA or cDNA) have been developed, and nested PCR is considered to be more sensitive than virus isolation, particularly since some enterovirus strains do not grow or fail to show CPE in cell culture.

Samples should be taken in the early phase of the disease since patients will usually excrete virus in the faeces for about 1 week (several weeks for children). Presence of virus is a strong indication of a causal relationship to disease. As virus shedding may be intermittent during the later phases of illness, a negative result does not exclude recent infection.

Antibody investigations. A test for specific IgM is used in some laboratories and is considered to be the method of choice for coxsackievirus B infections. The CFT is easy to perform, but because of the occurrence of cross-reactions the CFT is of limited value for enterovirus diagnosis.

MIGHT AS WELL TAKE SOMETHING ENJOYABLE

9. RHINOVIRUSES AND CORONAVIRUSES

Lat. *rhinus* = nose; Ger. *Erkältung*; Fr. *rhûme* = common cold.

I. Ørstavik

Rhinoviruses are the most frequent cause of common colds. All age groups are affected. Infections are endemic with higher frequencies during autumn and spring in temperate climates.

TRANSMISSION/INCUBATION PERIOD/CLINICAL FEATURES

The common cold is spread by close contact and by inhalation of virus-containing droplets. The incubation period is 2–4 days, and a person is probably infectious from 1 day postinfection and as long as there are clinical symptoms.

SYMPTOMS AND SIGNS

Systemic:	None or Low-Grade Fever, Headache
Local:	Coryza, Sneezing, Sore Throat, Cough, Hoarseness

In uncomplicated cases the illness usually lasts for 1 week, with maximal symptoms on days 2 and 3.

COMPLICATIONS

Secondary bacterial infections may occur (sinusitis, otitis media). Rhinovirus infections may precipitate acute asthma in predisposed children, and may aggravate chronic bronchitis in adults.

THERAPY AND PROPHYLAXIS

No specific therapy or prophylaxis is available.

LABORATORY DIAGNOSIS

During acute illness the virus can be isolated from the nose, the throat and sputum. Special cell culture techniques are needed for virus isolation and these are performed in very few virus laboratories. Serological diagnosis is not routinely used either, because of the many serotypes.

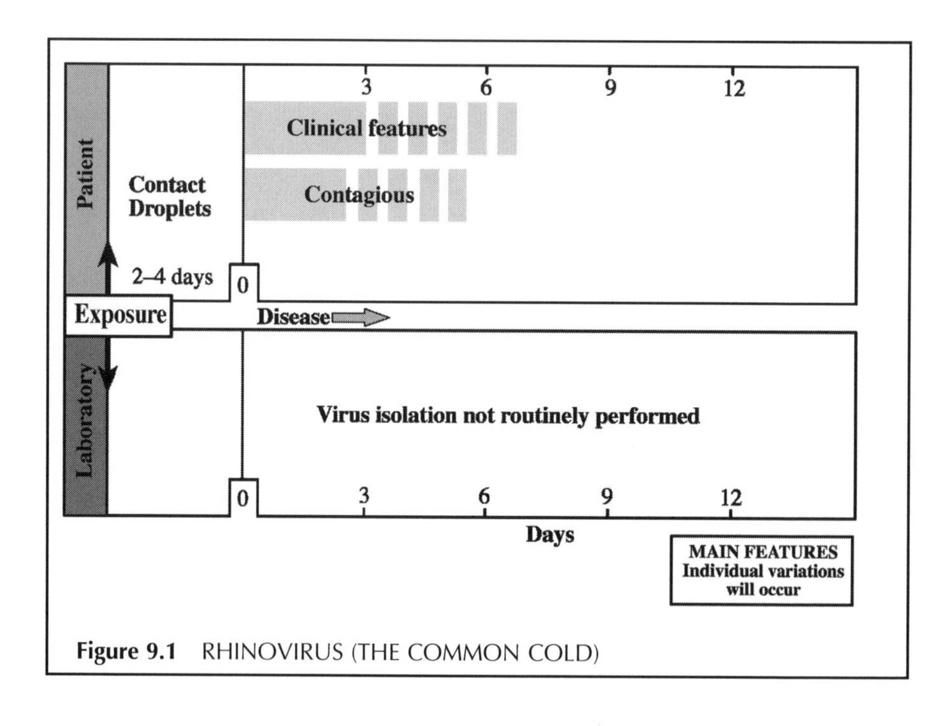

Figure 9.1 RHINOVIRUS (THE COMMON COLD)

CLINICAL FEATURES

SYMPTOMS AND SIGNS

After an incubation period of 2–4 days, the illness starts with symptoms of nasal congestion/blockage and irritation, sneezing and a sore throat. Excess nasal secretion follows which is serous at first and later becomes purulent if secondary bacterial infection ensues. Cough is a frequent symptom, as is headache during the first days of illness. Fever occurs seldom, and if so, it is moderate. Rhinovirus infection causes the same symptoms in all age groups. The infection is limited to the respiratory tract. It has been suggested that rhinoviruses may cause a more serious infection of the lower respiratory tract in small children. Rhinovirus infection has also been shown to precipitate attacks of asthma in children and aggravate chronic bronchitis in adults. Asymptomatic infections are reported to occur in about 25% of individuals infected with rhinovirus.

Differential diagnosis. Symptoms of common cold, particularly in children, may be due to other virus infections, e.g. influenzavirus, parainfluenzavirus, adenovirus, RSV and coronaviruses. Coronaviruses are now considered to be second to rhinoviruses as a cause of common cold, but the symptoms are usually milder in coronavirus infections. Influenzavirus infections occur in epidemics, and general symptoms such as fever and malaise are more severe. In parainfluenza and adenovirus infections pharyngitis is more pronounced. During epidemics of RSV some of the patients, children as well as adults, may have the same symptoms as in rhinovirus infections. Pharyngitis and tonsillitis will dominate infections with *Streptococcus pyogenes*. However, it is usually not possible to determine the aetiology on the basis of the clinical findings alone in upper respiratory infections.

CLINICAL COURSE

As a rule, the illness will last for 1 week, but 25% of the patients will need 2 weeks to recover completely. The illness tends to last longer in smokers than in non-smokers.

COMPLICATIONS

Bacterial sinusitis and otitis media are the most common complications. Occasionally a bacterial bronchopneumonia is seen.

THE VIRUS

Rhinovirus and *Enterovirus* are two genera in the family *Picornaviridae*. They are small (28–32 nm) single-stranded RNA viruses (Figure 9.2).

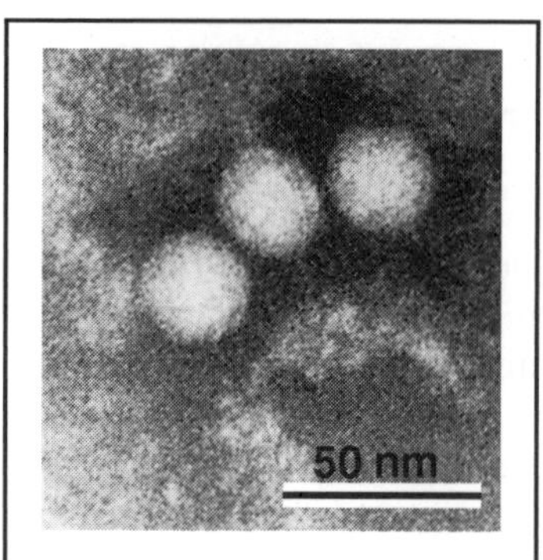

Figure 9.2 RHINOVIRUS (Electron micrograph courtesy of E. Kjeldsberg)

Rhinovirus now comprises more than 100 different serotypes, and new types are still being identified. As with other picornaviruses the virion capsid consists of a naked icosahedron of 60 capsomers, each made up of four proteins. Depressions in the virus capsid represent the sites on the virus where the cellular receptors bind. These depressions ('sockets') are the targets for experimental studies of synthetic antirhinoviral agents. A fifth protein is associated with the single-stranded RNA. Due to the lack of a lipid envelope, the virus is resistant to inactivation by organic solvents. Rhinoviruses are more acid-labile than enteroviruses.

EPIDEMIOLOGY

The rhinoviruses probably cause about half of all cases of common cold and are considered to be one of the most frequent causes of infections in man. Studies in the USA have revealed an infection rate of at least 0.6 per individual per year. The rate is highest among infants and decreases with age. Schoolchildren are considered to be important transmitters of rhinovirus infections. Parents with children in kindergarten or in primary school may have more common cold episodes than single adults. Rhinovirus infections are endemic, but occur most frequently during autumn and spring in temperate climates. Several serotypes can circulate simultaneously in the same population, and it is possible that new serotypes emerge over the years. There is no evidence that some serotypes cause more serious illness or occur more frequently than others.

THERAPY AND PROPHYLAXIS

Specific chemotherapy is not available, and treatment with immunoglobulin is without effect. Experiments in volunteers have found α- and β-interferon given intranasally to be effective in preventing rhinovirus infection, whereas studies using γ-interferon have been unsuccessful. The suggestion that large quantities of vitamin C (ascorbic acid) taken prophylactically or during illness influences the course of the disease, has not been proven. Symptomatic treatment includes mild analgesics and nasal drops. Prophylactic use of antibiotics against bacterial superinfections is not recommended in otherwise healthy individuals.

No **vaccine** is available. The high and uncertain number of serotypes and their relative importance and distribution during various outbreaks are obstacles to vaccine development. In addition to inhalation of droplets, spread of infection by contact is considered to play a significant role. Measures should be taken to avoid infection from virus-contaminated hands. Persons suffering from asthma and from chronic bronchitis should avoid close contact with common cold patients.

LABORATORY DIAGNOSIS

Cultivation of rhinoviruses requires special cell cultures which are incubated at 33°C (the temperature in the nasal mucosa). Also, since many serotypes are difficult to cultivate, rhinovirus isolation is performed only by very few virus laboratories. Serological diagnosis is complicated by the large number of serotypes and is therefore not routinely performed.

CORONAVIRUS

Coronaviruses are the second most frequent cause of the common cold (15–20%). They are single-stranded RNA viruses belonging to the *Coronaviridae* family. The virions vary in diameter from 80 to 160 nm. They have club-shaped spikes on the surface which give a crown (corona)-like picture by electron microscopy (Figure 9.3). At least four different proteins are known, and the S (spike)-protein induces virus-neutralizing antibodies contributing to immunity. The coronaviruses are divided into three serological groups, the human coronaviruses have been allocated to two of these serological groups, and the two human prototypes are OC43 and 229E. The coronaviruses are believed to spread as the rhinoviruses, and the incubation period is about 2 days. The symptoms are similar to those following rhinovirus infections, lasting for about 1 week. As many as 50% of coronavirus infections may be asymptomatic. Serological studies suggest that the infection occurs in all age groups. Reinfection viruses have been observed, suggesting that protective immunity is not long-lasting. Coronavirus infections occur most frequently in late winter/early spring. Coronavirus may be isolated from the nose and throat during the acute phase of illness if organ cultures of human fetal trachea are used. Only a small number of coronavirus strains have been identified, and most knowledge about this virus infection has been obtained by serological studies on paired sera from patients. Very few laboratories diagnose coronavirus infections as part of their routine work.

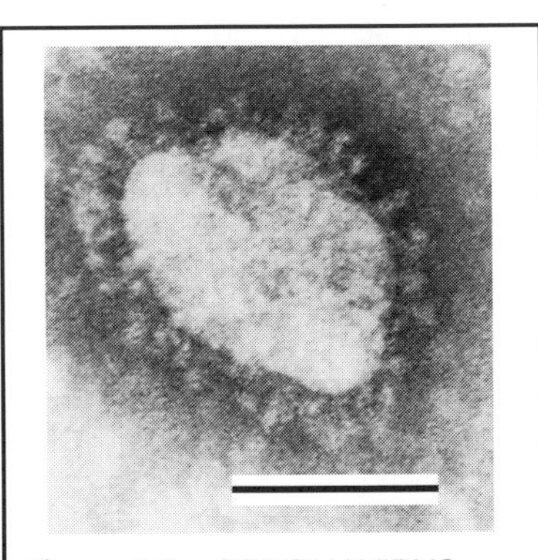

Figure 9.3 CORONAVIRUS. Bar, 100 nm (Electron micrograph courtesy of E. Kjeldsberg)

A REAL KNOCK-OUT

10. INFLUENZAVIRUSES

Influenza; influenced by cosmic events (medieval Italy). Ger. *Grippe*; Fr. *grippe*.

L. R. Haaheim

Influenzavirus causes illness in all age groups. During epidemics a large number of individuals may fall ill within a span of a few weeks.

TRANSMISSION/INCUBATION PERIOD/CLINICAL FEATURES

Virus is transmitted by aerosols, mean incubation time is 2 days (1–4). The patient is contagious during the first 3–5 days of illness.

SYMPTOMS AND SIGNS

Systemic:	Sudden Fever (38–40°C), Myalgia, Headache
Local:	Coryza, Dry Cough, Sore Throat, Hoarseness

Systemic symptoms dominate initially with fever for the first 3–4 days. Full recovery within 7–10 days. Occasionally long convalesence.

COMPLICATIONS

Secondary bacterial pneumonia. More rarely primary viral pneumonia, myocarditis, encephalitis.

THERAPY AND PROPHYLAXIS

No specific treatment. Amantadine chemoprophylaxis. Vaccination is recommended for high-risk groups and key personnel within the health services.

LABORATORY DIAGNOSIS

Virus can be isolated from/demonstrated in nasopharyngeal specimens taken in the acute phase of illness (days 1–3). An antibody rise can be demonstrated in paired sera by HI or CFT.

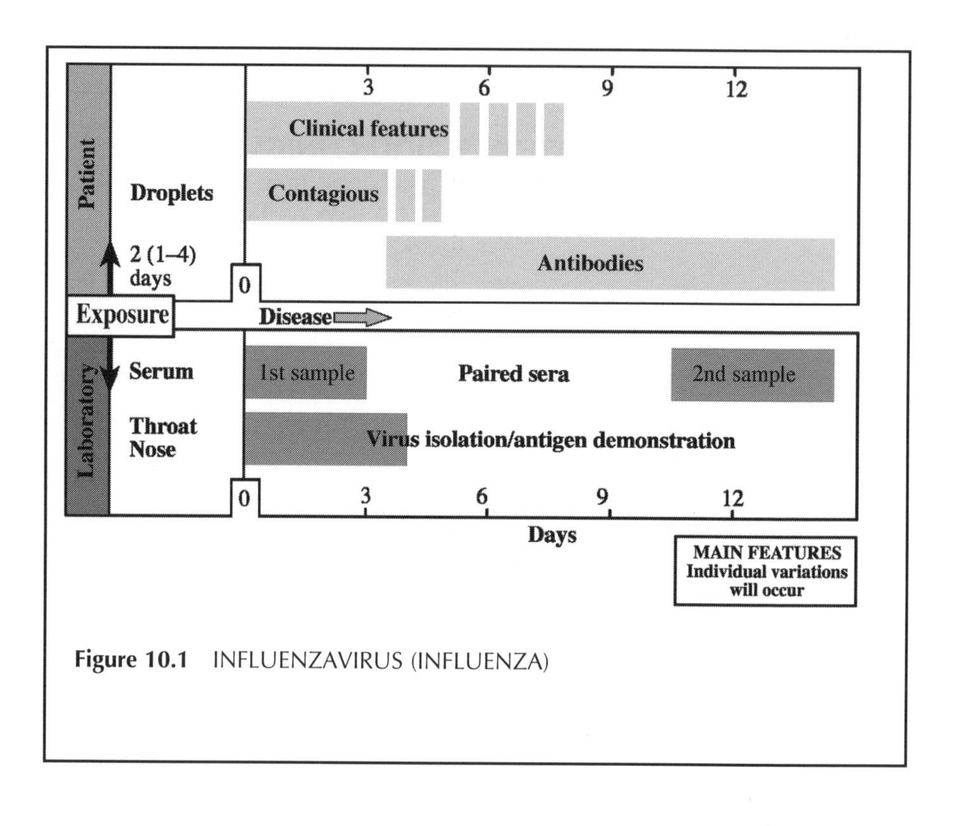

Figure 10.1 INFLUENZAVIRUS (INFLUENZA)

CLINICAL FEATURES

SYMPTOMS AND SIGNS

Virus is transmitted by aerosols. The mean **incubation time** is 2 days (1–4). Prodromal symptoms are uncommon. A sudden onset is most typical, with chills and rising fever followed by myalgia (usually back pain in adults). These initial systemic manifestations may later be followed by signs of pharyngitis/laryngitis/tracheobronchitis. In conjunction with dry cough the patient may complain of substernal pain. Other symptoms may be coryza, flushed face, epistaxis, photophobia, anorexia or vertigo. Croup among small children is a relatively common feature (see also Chapter 11).

Differential diagnosis. A range of acute respiratory diseases are often mistaken for 'influenza'. Only laboratory tests can establish the viral aetiology.

CLINICAL COURSE

The fever, after having peaked at about 38–40°C within 1–3 days, may fall abruptly, but usually there is a gradual defervescence in the course of 2–4 days. Occasionally a second period of fever occurs. When afebrile the patient may have a productive cough and suffer from fatigue. Respiratory tract symptoms usually dominate the later phases of illness. Damage to the mucociliary epithelium in the airways may take weeks to repair. Otherwise healthy patients usually recover completely within 7–10 days.

COMPLICATIONS

The most common complication is secondary bacterial pneumonia which is especially serious among the elderly and those with chronic disease of the heart, lungs and bronchi. The excess mortality associated with some influenza epidemics is a serious community health problem. A biphasic fever may indicate bacterial superinfection, and the bacterial pneumonia has a particularly rapid course. Primary viral pneumonia is a rarer event and may be recognized by a prolonged primary fever, dyspnoea, hypoxia and cyanosis. Encephalitis/encephalopathy is an extremely rare complication. Postinfectious Reye's syndrome has sometimes been seen in connection with influenza, especially that due to influenza B virus. Myocarditis and transient cardiac arrhythmia have occasionally been observed.

THE VIRUS

Influenzavirus is an enveloped single-stranded RNA virus belonging to the family *Orthomyxoviridae* (Figure 10.2). Three types of influenzavirus are known, namely types A, B (genus *Influenzavirus A, B*) and C (genus *Influenzavirus C*). Types B and C are believed to have man as the only host, whereas type A virus is found in a wide range of species, e.g. swine, horses and birds. Two different types of spikes, the haemagglutinin (H) and the neuraminidase (N) protrude from the viral envelope. The former is responsible for attachment of virus to cellular receptors, and anti-H antibodies protect against infection. The neuraminidase is believed to play a role in virus release from infected cells. The neuraminidase splits off neuraminic acid from the cellular virus receptor. Only rarely will anti-N antibodies neutralize viral infectivity. For influenza A the internal proteins are the matrix antigen (M1) surrounding the nucleocapsid and closely attached to the viral membrane, the nucleoprotein (NP) associated with the segmented viral genome and the polymerases PA, PB1 and PB2. The division into types is based on the antigenic properties of the M1 and NP proteins. Another protein (M2), coded for by the same gene segment as M1, is present in small quantities in the infectious virion and functions as a proton channel. Both surface antigens show a marked tendency for antigenic variation ('drift'). This is the main reason for the frequent recurrence of epidemics. Based on genetic and antigenic characteristics of influenza A surface antigens they can be divided into subtypes, of which there are 15 for the haemagglutinin and 9 for the neuraminidase. The segmented viral genome allows for formation of viral reassortants ('recombinants') between different strains or subtypes of virus. A doubly-infected host can thus give rise to a 'new' virus. Such a profound change in antigenic make-up ('shift') is probably the mechanism for the occurrence of new pandemic A strains. Exchange of genes between animal and human influenza A subtypes may play an important part in this phenomenon. Also gene combinations for virulence can be reassorted between viruses. In Hong Kong in 1997 a very virulent avian influenza strain (A/H5N1) surprisingly infected people directly without any animal intermediary. No interhuman spread was observed. This zoonosis caused considerable international concern.

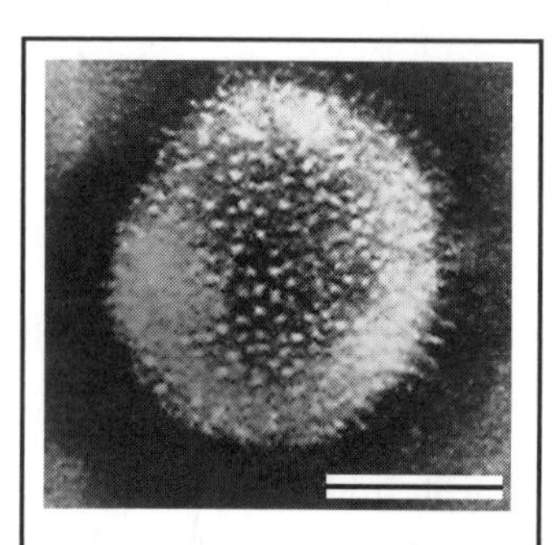

Figure 10.2 INFLUENZAVIRUS. Bar, 50 nm (Electron micrograph courtesy of G. Haukenes)

EPIDEMIOLOGY

New influenza strains with changes in one or both surface antigens are able to evade the herd immunity built up to previous strains. In 1957 the A/H2N2 subtype ('Asian flu') replaced the A/H1N1 virus, resulting in a pandemic situation. In 1968 a new shift occurred giving rise to the A/H3N2 subtype

('Hong Kong influenza'). Since 1968 variants of the H3N2 virus have continued to circulate. Unexpectedly, in 1977 the A/H1N1 subtype reappeared in China after being out of circulation since 1956. Initially this virus predominately infected young individuals and spread all over the world in the course of approximately 1 year. Variants of this virus have continued to circulate. Thus, presently we have strains of two influenza A subtypes (H3N2 and H1N1) co-circulating. Influenza A pandemics occur at long and unpredictable intervals (decades), whereas larger epidemics of influenza A appear on average every 2–4 years. Influenza B virus is less prone to antigenic drift and gives larger outbreaks somewhat less frequently. In the temperate zones epidemics usually occur during the winter season, but sporadic cases/outbreaks may take place throughout the year. Factors determining whether epidemics take place or not are not well understood. The antigenic properties of the virus together with factors related to transmissibility and virulence and the extent of herd immunity in the population all play a role. Age-related as well as regional and local differences in severity and degree of epidemic impact are frequently observed. Thus, prediction of epidemics and their special parameters is extremely difficult and in most cases not possible. The World Health Organization (WHO) has set up a global surveillance system for influenza. About 110 national institutes in more than 80 countries form a cooperative international network whose main task is to monitor epidemics, isolate strains and disseminate information to all members on a weekly basis throughout the winter season. WHO recommendations for the antigenic composition of the vaccines are based on an international scientific consensus.

THERAPY AND PROPHYLAXIS

Treatment is mostly symptomatic. Extensive use of salicylates should be discouraged due to its possible precipitating effect on the development of Reye's syndrome. **Immunoglobulin** has neither therapeutic nor prophylactic effect. Amantadine and its derivatives have a proven **prophylactic** and **therapeutic** effect on influenza A (see p. 31). New promising drugs against both influenza A and B, targeted against the virion neuraminidase activity, are under clinical trials. **Influenza vaccine** will, provided the antigenic formulation does not differ significantly from the epidemic strain, confer protection in about 75% of recipients. The inactivated vaccines used throughout the world are purified egg-grown virus killed by formalin or by β-propriolactone. The use of live attenuated vaccine is limited. Preparations containing detergent split virus or isolated surface subunits (H and N) are mostly used. The two latter types of vaccine are claimed to give fewer side-effects than whole virus vaccine. About 15–40% of the vaccinees will experience slight redness and induration at the injection site, and some show signs of mild influenzal disease of short duration. Apart from the association of polyradiculitis with the swine influenza vaccine programme in the USA in 1976, there are no known systemic complications to influenza inoculation. Persons for whom clinical influenza would lead to further deterioration of their underlying condition are recommended as target groups for routine vaccination. Generally, these are the elderly, those with chronic illnesses in the heart, lungs and airways, and those with metabolic disorders or immune deficiencies. As a routine measure, influenza vaccination should be carried out before the usual time of onset of epidemics. In temperate climates this means vaccination during the autumn months. Protective levels of anti-influenza antibody will ensue 1–3 weeks postvaccination. Immunity to influenza, whether through natural infection or vaccination, will last for about a year as the ever-changing virus will outdate acquired immunity.

LABORATORY DIAGNOSIS

Influenzavirus can be isolated in cell cultures or in fertile hens' eggs from nasopharyngeal samples taken in the acute phase of illness. The physician may get a laboratory answer within a week. More rapid diagnostic methods, demonstrating (e.g. by IF) virus or viral material in the sample within a few hours, are now available in some laboratories. More specialized laboratories offer subtype identification of influenza A virus. In paired sera, taken in the acute phase and 10–14 days after onset, a significant CF antibody rise is indicative of recent infection with either A or B virus. HI tests may in some cases indicate which influenza A subtype is the aetiological agent. However, HI tests may in some instances give misleading or inconclusive results due to 'original antigenic sin'. This phenomenon is a poorly understood anamnestic HI antibody recall of childhood influenza memory triggered by current heterotypic strains.

HOME CURED

11. PARAINFLUENZAVIRUSES

The name relates to the affinity of the virus for the respiratory tract giving mild influenza-like diseases.

A. B. Dalen

Parainfluenzaviruses (four serotypes) are important pathogens of the respiratory tract in infants, children and young adults. They are the major cause of croup, and also cause bronchiolitis and pneumonia.

TRANSMISSION/INCUBATION PERIOD/CLINICAL FEATURES

Virus is transmitted by close contact or inhalation of droplets. The incubation period is 2–4 days (children) and 3–6 days (adults).

SYMPTOMS AND SIGNS

Systemic:	Slight Fever and Malaise
Local:	Cough, Hoarseness, Coryza, Croup

Most children recover within 3–6 days.

COMPLICATIONS

Rarely seen. Atelectasis may develop following pneumonia.

THERAPY AND PROPHYLAXIS

No specific antiviral treatment. Symptomatic treatment aims to relieve respiratory distress. The use of corticosteroids is controversial.

LABORATORY DIAGNOSIS

Viruses may be isolated in cell cultures. PCR detection may be used. IF is used to demonstrate viral antigen in nasopharyngeal aspirates. Serodiagnosis is of little practical value.

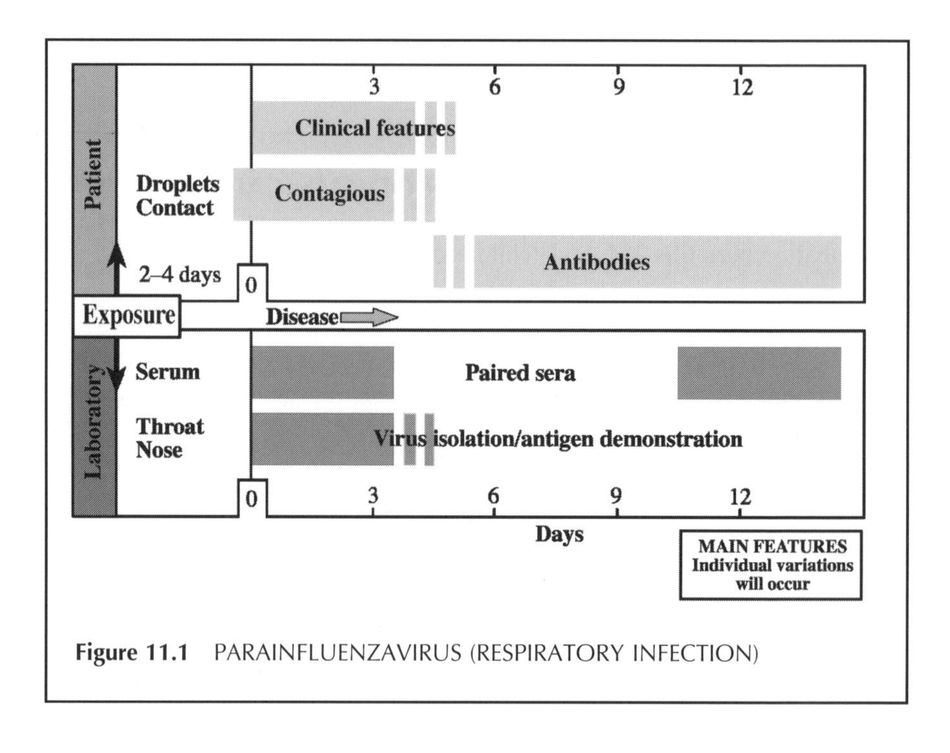

Figure 11.1 PARAINFLUENZAVIRUS (RESPIRATORY INFECTION)

CLINICAL FEATURES

SYMPTOMS AND SIGNS

The incubation period is 2–4 days. After signs of rhinitis and pharyngitis for a few days, the patient may become hoarse and have inspiratory stridor. There may be a mild to moderate fever. Involvement of the lower respiratory tract (bronchitis and bronchopneumonia) is seen in 30% of primary infections with type 3 virus. Severe croup is seen in 2–3% of primary infections with virus types 1 and 2. The parainfluenzaviruses are second only to respiratory syncytial virus as the cause of serious respiratory tract infections in infants and children. Parainfluenzavirus 1 and 2 most often give infections affecting the larynx and the upper trachea which may result in croup, while type 3 has a predilection for the lower respiratory tract giving bronchitis, bronchiolitis and bronchopneumonia. Type 4 virus is less virulent, being associated with mild upper respiratory tract illness in children and adults. Due to the presence of protective maternal antibodies, serious illness due to parainfluenzavirus 1 and 2 infections is not seen before the age of 4 months. For type 3 virus serious illness is seen in the first months of life in spite of the presence of maternal neutralizing antibodies. The incidence of severe infections increases rapidly after the age of 4 months, peaking between the age of 3 and 5 years. The incidence is lower when the child reaches school age, and clinical disease from parainfluenzavirus 1, 2 and 3 infections is unusual in adult life. Reinfections with the same type of virus occur frequently, but with milder clinical manifestations.

Differential diagnosis. Croup is occasionally caused by RSV and influenzaviruses. In bronchitis, bronchiolitis and bronchopneumonia in infancy other viruses, notably RSV, must be considered. *Chlamydia trachomatis* causes lower respiratory tract infections in early infancy, while *Chlamydia pneumoniae* and *Mycoplasma pneumoniae* are more commonly found in older children and young adults. The most important differential diagnosis to viral croup is bacterial croup or epiglottitis caused by *Haemophilus influenzae* type B. This is a life-threatening condition which usually starts without prodromal rhinitis and hoarseness. The patient has dysphagia and a higher fever than in viral croup and a toxic appearance. The epiglottis appears enlarged and inflamed on inspection. Respiratory distress does not tend to be diminished by bringing the patient into an upright position. Diphtheritic croup is now a rare illness in many countries. The condition is characterized by marked swelling of tonsils, the presence of membranes, prostration and high fever.

CLINICAL COURSE

Fever usually lasts for 2–3 days. Most children recover uneventfully from croup after 24 to 48 hours. When bronchiolitis and pneumonia develop, fever and cough persist for some time.

COMPLICATIONS

Atelectasis may develop following lower respiratory tract infections. Complications are otherwise very rare.

THE VIRUS

The parainfluenzaviruses belong to the genus *Paramyxovirus*, family *Paramyxoviridae* (Figure 11.2). The viral genome, a single-stranded RNA of negative polarity and about 15,500 nucleotides in length, is surrounded by core proteins and an outer lipid membrane, bearing glycoprotein spikes. These surface glycoproteins include a bifunctional protein, the haemagglutinin–neuraminidase, and the fusion protein. The haemagglutinin is responsible for attaching the virus to host-cell receptors. The neuraminidase functions late in the infection cycle, releasing new virions from infected cells. The fusion protein mediates penetration of the viral core through the cytoplasmic membrane of the host cell. This protein also mediates the characteristic fusion of cells seen in infected tissue. The fusion protein is rendered biologically active by a cellular protease, a process which is essential for infectivity. Antibodies against the haemagglutinin–neuraminidase and the fusion proteins have neutralizing properties. The parainfluenzaviruses share antigens with mumps virus and parainfluenzaviruses of animal origin. The four serotypes of parainfluenzavirus differ in antigenic composition and to some extent in cytopathogenicity and clinical manifestations.

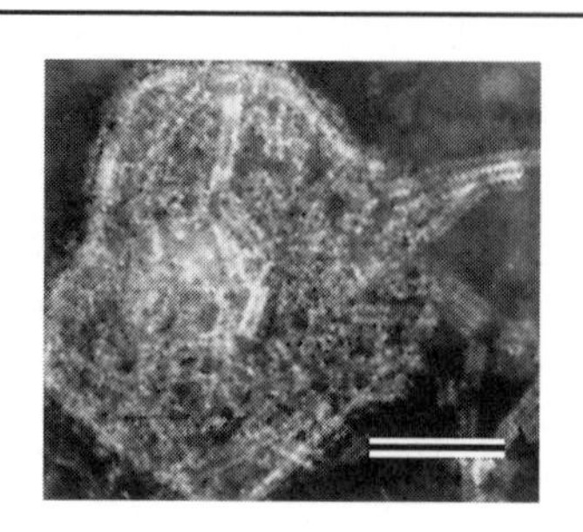

Figure 11.2 PARAINFLUENZAVIRUS: RUPTURED VIRION WITH HELICAL NUCLEOCAPSIDS. Bar, 100 nm (Electron micrograph courtesy of G. Haukenes)

EPIDEMIOLOGY

Parainfluenzaviruses 1 to 4 are ubiquitous and infect the respiratory tract. There is some seasonal variation with fewer cases during the summer months. The viruses are readily spread, which gives rise to high infection rates in early life. Type 3 virus is more easily spread than types 1, 2 and 4, and occurs more commonly during infancy and early childhood. Infections with type 1 and 2 virus occur somewhat later, but the majority of children have experienced infections with these viruses by the age of 5. Infections give rise to both local nasal secretory IgA and serum-neutralizing antibodies. The IgA from adults neutralizes viral infectivity, while locally produced IgA from young infants lacks or has little neutralizing capacity. Serum-neutralizing antibodies are only partially protective, and reinfections probably occur repeatedly. The clinical manifestations are generally less severe during reinfections. After the age of 7, parainfluenzavirus infections are usually subclinical.

THERAPY AND PROPHYLAXIS

Inactivated **vaccines** have been shown not to prevent parainfluenzavirus infection or disease. Specific antiviral **treatment** is not available. The importance of interferon in recovery from parainfluenzavirus disease is not known. Symptomatic treatment of croup includes keeping the patient in an upright position in a humidified and cooled atmosphere, and correction of hydration. Regimens including the use of nebulized racemic epinephrine and systemic corticosteroids are controversial. Intubation may be indicated on rare occasions with severe respiratory distress.

LABORATORY DIAGNOSIS

Routine laboratory diagnostics are usually limited to parainfluenzaviruses 1, 2 and 3. The viruses grow well in tissue culture. The lability of the viruses, however, makes inactivation of the virus during transportation a problem. PCR detection avoids this problem. The direct demonstration of virus-infected cells from nasopharyngeal aspirates by immunofluorescence is a very good practical alternative. A conclusive answer can be given within a few hours. Serodiagnosis by HI, CF or NT requires paired sera taken 1–3 weeks apart. This delay renders serodiagnosis impractical in a clinical setting.

OF INFLATED IMPORTANCE?

12. MUMPS VIRUS

Epidemic parotitis. Ger. *Ziegenpeter*; Fr. *oreillon.*

B. Bjorvatn and G. Haukenes

Inflammation of the salivary glands may be caused by bacterial, fungal or viral infection, or by toxic–allergic reactions. Acute enlargement of the salivary glands in children and young adults is mostly due to infection with mumps virus.

TRANSMISSION/INCUBATION PERIOD/CLINICAL FEATURES

Infection is transmitted through inhalation of virus-containing aerosols. The incubation period is usually 2–3 weeks. Period of communicability is from a few days before, to about 1 week after clinical onset.

SYMPTOMS AND SIGNS

Systemic:	Fever
Local:	Painful Swelling of Salivary Glands
Others:	*See* Complications

In uncomplicated cases recovery is complete within 1 week.

COMPLICATIONS

Most common complications are meningitis and orchitis. The prognosis is usually good.

THERAPY AND PROPHYLAXIS

No therapeutic or prophylactic value of drugs or specific immunoglobulin. Complications are treated according to symptoms. Live attenuated virus vaccines are available that provide more than 90% protection.

LABORATORY DIAGNOSIS

Virus can be isolated in cell culture from saliva collected during the first week of disease, and in urine somewhat longer. During the first 4–5 days of mumps meningitis, the virus may be found in spinal fluid. Serum antibodies are detectable 1–3 weeks after clinical onset. The serological diagnosis is based on seroconversion or the specific IgM by ELISA.

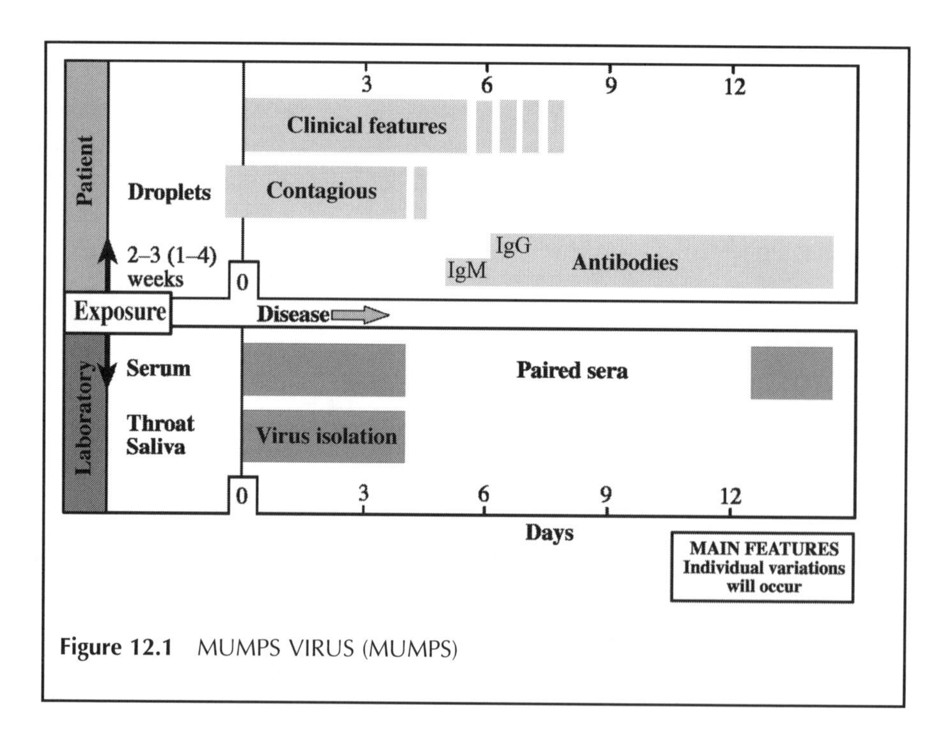

Figure 12.1 MUMPS VIRUS (MUMPS)

CLINICAL FEATURES

SYMPTOMS AND SIGNS

The incubation period is usually 2–3 weeks. Typically, the patient develops slight/moderate fever some days before swelling of the salivary glands, mostly the parotid glands. In about 80% of the cases there is a bilateral swelling, appearing within an interval of one or several days. A continuous swelling involving the salivary glands, the jaw region and lateral aspects of the neck is not uncommon. The patient complains of oral dryness and painful chewing, and occasionally there is trismus. Oedema and redness around the opening of the Stensen's duct are frequently seen. Mumps that remains located symptomatically to the salivary glands is considered uncomplicated and recovery is usually complete within 1 week or less. Asymptomatic infections are common (at least 20–30%), particularly in early childhood.

Differential diagnosis. Mononucleosis or bacterial infections of oropharynx, sialoadenitis (anomalies of the glandular duct, immune deficiency), other viral infections of salivary glands (rare), allergic reactions, collagen disease and lymphoma. The diagnosis is usually made clinically, particularly during epidemics. In case of doubt, laboratory confirmation should be obtained.

CLINICAL COURSE

In uncomplicated cases of mumps complete recovery is expected in 1 week or less.

COMPLICATIONS

Asymptomatic pleocytosis (>5 leukocytes/mm^3) of the cerebrospinal fluid is found in 50–60% of mumps cases. **Symptomatic meningitis or meningo-encephalitis** occurs in 1–3% of cases, three times more frequently in males than in females. Clinically, the picture is dominated by headache, rigidity of neck, nausea, emesis and fever. Examination of spinal fluid often reveals a considerable mononuclear pleocytosis. There is poor correlation between severity of clinical disease and number of cells in the cerebrospinal fluid. **Meningitis** occurs at the time of, or 4–7 days following glandular involvement, but may occur in the absence of salivary gland swelling. The prognosis is good. Young children tend to recover in a few days, whereas teenagers and particularly adults occasionally require weeks (months) for complete recovery. Mumps encephalitis without signs of meningitis is reported in 0.02–0.3% of cases. Permanent neurological sequelae such as unilateral hearing loss or facial nerve palsy may occur in such cases.

Orchitis complicates the course in approximately 20–30% of postpubertal males. This condition is characterized by fever, intense pain and often considerable swelling of the testicle, frequently occurring when the salivary gland enlargement has subsided. Orchitis is usually unilateral, but the other testicle may occasionally (20%) be affected within a few days. Prostatitis and epididymitis may also occur. The symptoms of mumps orchitis normally disappear within 1 week. Although parenchymatous necrosis may lead to some testicular atrophy, permanent infertility is very rare, even in cases of bilateral involvement. Temporary reduction in fertility is not uncommon, however. A history of mumps orchitis appears to be a risk factor for testicular cancer.

Oophoritis occurs in about 5% of female mumps patients, but does not cause infertility.

Pancreatitis, diagnosed in about 7% of cases, heals uneventfully. Note that high levels of amylase may result from swelling of the salivary glands only. Occasionally other glands (thyroid, lacrimal, mammary and thymus) are involved. On rare occasions involvement of liver, spleen, joints, myocardium, retina and conjunctiva has been reported.

THE VIRUS

Mumps virus (Figure 12.2) belongs to the genus *Paramyxovirus* of the family *Paramyxoviridae*. The mumps virus particle is pleomorphic, 150–200 nm in diameter. The envelope and the underlying matrix surround a helical nucleocapsid containing a single molecule of negative-sense single-stranded RNA. The envelope contains three glycosylated proteins, H (haemagglutinin, attachment protein), F (fusion protein, haemolysin) and N (neuraminidase). The M (matrix) protein lies beneath the envelope. Associated with the nucleocapsid are P (phosphoprotein) N and NP (RNA-binding protein) and a large (L) protein (RNA polymerase). Virus enters the cell by fusion with the cell membrane. The viral RNA is transcribed to a positive strand from which proteins are translated and new genomic RNA is made. During this transcription two additional proteins are coded by the P gene, one by ribosomal frameshift (C protein) and one (V protein) by a newly recognized strategy including occasional insertion of additional nucleotide resulting in frameshifting (RNA editing). Nucleocapsids are made in the cytoplasm, and the virus matures by budding from the cell membrane. The virus is unstable and easily destroyed by ether, heating at 56°C for 20 minutes, ultraviolet irradiation and treatment with most disinfectants.

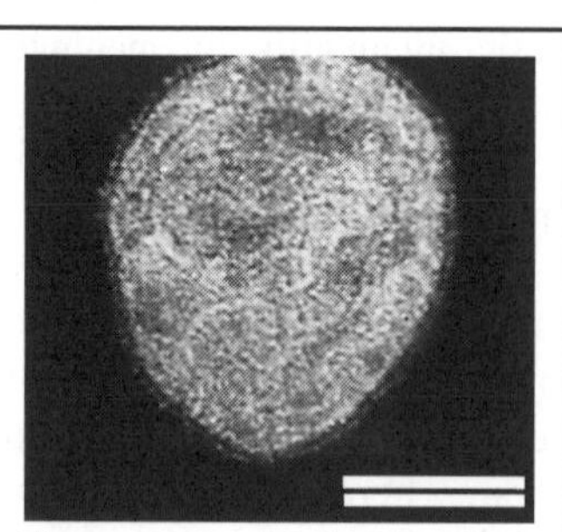

Figure 16.2 MUMPS VIRUS. Bar, 100 nm (Electron micrograph courtesy of G. Haukenes)

EPIDEMIOLOGY

Mumps virus is mainly transmitted by airborne droplets when contagious persons are in contact with susceptible individuals. Although mumps virus may be found in the saliva from 6–7 days before and up to 8 days following onset of disease, transmission is most efficient around the time of onset. Transmission also occurs in cases of subclinical infection. The susceptibility is considerable in non-immune populations, as reflected by annual incidence rates between 0.1 and 1%. Peak incidences are found in the 5–7 year age group. Small children, and in particular infants, rarely contract the disease, or the disease more often runs an asymptomatic course. In tropical climates mumps is endemic throughout the year. In temperate climates the disease is most prevalent in the winter and spring, tending to cause larger epidemics at 3–5 year intervals. With increasing vaccination coverage, mumps has largely disappeared in many industrialized countries.

THERAPY AND PROPHYLAXIS

There is no specific chemotherapy available, and high-titred **immunoglobulin** has no proven therapeutic effect. In cases of meningitis the patient should remain in bed and analgesics, antipyretics and antiemetics should be provided as needed. Similar symptomatic treatment is instituted for orchitis, which in addition may require a mechanical support, local cooling (ice bags) and possibly systemic corticosteroids. Prophylactic use of specific immunoglobulin has no documented effect. Safe and effective live attenuated **vaccine** based on virus from chick embryo fibroblasts provides more than 90% protection for at least 10 years following one single injection. Mild side-effects (low-grade fever, local tenderness) are occasionally seen. On very rare occasions such vaccine strains may cause a mild meningitis. Mumps vaccine, either as a monovalent or a combined trivalent vaccine with measles and rubella (MMR), may be offered at the age of 15–24 months, followed by a second injection at prepuberty. Pregnancy, ongoing infectious diseases (except mild infections) and immune deficiency are all contraindications for such live vaccines.

LABORATORY DIAGNOSIS

During the first week of disease, virus can be isolated from saliva or oral washings. Successful virus isolation may be made from urine for another 1–2 weeks. Virus may also be recovered from the cerebrospinal fluid during the acute stage of meningoencephalitis. Mumps virus replicates in different types of cell cultures, but monkey kidney cell lines are preferred. The virus-induced cytopathic effect manifests as giant cells and degenerative changes leading to cell death. Virus-infected cells adsorb erythrocytes, and this haemadsorption (Had) is inhibited by specific mumps antibodies (HadI). Virus is identified by neutralization, immunofluorescence (cytoplasmic and membrane) or by HadI.

For serological diagnosis, seroconversion or specific IgM is usually looked for. Significant rises in titre are looked for by ELISA. Antibodies can be detected 1–3 weeks after clinical onset. Detection of specific IgM indicates onging or recent (within weeks/months) infection with mumps virus. Possible cross-reaction with other parainfluenzaviruses should be kept in mind in evaluation of immunity status.

RAPID DIAGNOSIS LEADS TO CORRECT MANAGEMENT

13. RESPIRATORY SYNCYTIAL VIRUS (RSV)

The name reflects the ability of the virus to induce syncytia (giant cells) in tissue cultures.

G. Ånestad

RSV is the most important pathogen encountered in lower respiratory tract infections (bronchiolitis and pneumonia) in infants and small children. Among older children and adults reinfections are common. Clinically, these reinfections are usually manifested as upper respiratory tract infection (URTI). Epidemics of RSV occur regularly during the colder months in temperate climates and during the rainy season in tropical areas.

TRANSMISSION/INCUBATION PERIOD/CLINICAL FEATURES

Infection is transmitted by contact with infectious material and by aerosol. The incubation period is 3–6 days. The patient is contagious 1–2 weeks after onset of symptoms.

SYMPTOMS AND SIGNS

Systemic:	Moderate Fever, Hypoxaemia, Fatigue
Local:	Coryza, Cough, Respiratory Distress
Other:	*See* Complications

The period of critical illness lasts a few days. Infants and small children often have a convalescent period of some weeks with cough and fatigue.

COMPLICATIONS

Bacterial superinfections are uncommon. Small children who develop bronchiolitis are probably predisposed to develop asthma in later life. Viral otitis media is seen in 20% of patients.

THERAPY AND PROPHYLAXIS

Ribavirin administered as an aerosol may have a beneficial effect. Otherwise only symptomatic treatment is available. In infants and small children RSV pneumonia and bronchiolitis may be life-threatening, requiring immediate hospitalization. RSV immunoglobulin may have some prophylactic effect. No vaccine is available.

LABORATORY DIAGNOSIS

Antigen detection (IF, ELISA) in exfoliated nasopharyngeal cells is widely used. Serological examinations for significant titre rise are often unrewarding, particularly in infants.

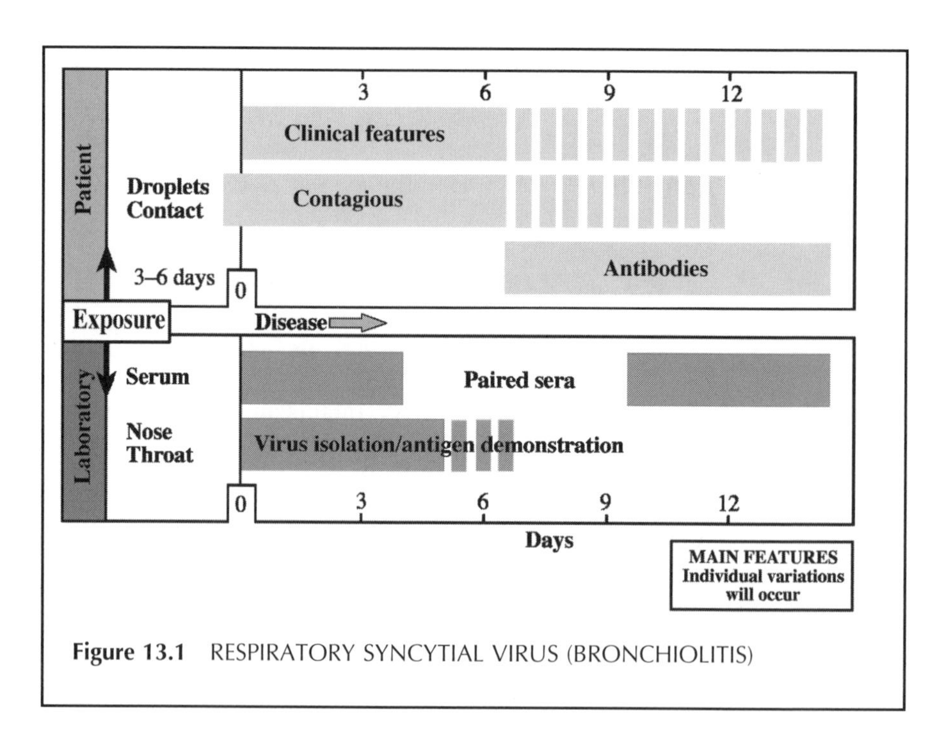

Figure 13.1 RESPIRATORY SYNCYTIAL VIRUS (BRONCHIOLITIS)

CLINICAL FEATURES

SYMPTOMS AND SIGNS

The period from exposure to onset of symptoms is usually 3–6 days. Among both small children and adults the first symptoms are those of URTI with cough and rhinorrhoea. RSV bronchiolitis develops only in small children, and the most severe cases are usually children under 12 months of age. Lower respiratory tract involvement usually occurs within the first week of illness. Clinically, wheezing and increased respiratory rate with intercostal and subcostal retractions are seen. In severe cases cyanosis, listlessness and apnoea may occur. RSV bronchiolitis and pneumonia are often difficult to differentiate, and many infants appear to have both. Chest radiographs may be normal, but often show a combination of air trapping (hyperexpansion) and bronchial thickening or interstitial pneumonia. Fever is seen in the URTI period and is moderate (38–40°C). Preterm infants and children with underlying diseases, in particular those with cardiopulmonary and congenital heart disease, are at high risk for developing severe RSV infection. Among school children and adults, RSV infection manifests as a common cold. Elderly persons, and in particular those in institutions, may develop pneumonia, sometimes with fatal outcome.

Differential diagnosis of RSV infection in infants and small children includes other causes of lower respiratory tract illness in this age group, in particular other respiratory virus infections (parainfluenzaviruses, influenzavirus, adenovirus). In infants infection with *Chlamydia trachomatis* may cause interstitial pneumonia with cough and in some instances wheezing. Contrary to the abrupt onset of RSV bronchiolitis, this latter illness tends to be subacute, and in approximately 50% of the cases the illness is heralded by a chlamydial conjunctivitis. In immunodeficient children, *Pneumocystis carinii* infection must be considered. In some infants with RSV infection, the cough may be so severe and paroxysmal that the illness may mimic the pertussis syndrome. The epidemic occurrence of RSV infection may be a guide to correct diagnosis.

CLINICAL COURSE

In hospitalized children the critical period usually lasts for 3–6 days. However, hypoxaemia may last for some weeks. The mechanisms involved in recovery of RSV infection are not fully understood. Contrary to other respiratory viruses, RSV induces little or undetectable levels of interferon and improvement usually coincides with the development of local and humoral antibodies.

COMPLICATIONS

Secondary bacterial complications are rare. Therefore, indiscriminate use of antibiotics should be discouraged. Approximately 20% of children with RSV bronchiolitis develop a viral otitis media. Evidence is now accumulating that small children who have had RSV bronchiolitis during their first year of life are predisposed to developing chronic respiratory disease, in particular asthma.

THE VIRUS

RSV (Figure 13.2) belongs to the genus *Pneumovirus* in the family *Paramyxoviridae*, subfamily *Pneumovirinae*. The linear single-stranded negative-strand RNA has a length of about 15 kb. Viral replication takes place in the cytoplasm of virus-infected cells and, like other members of the family *Paramyxoviridae*, infectious virions are released by budding through the cell membrane. However, RSV has no haemagglutinin or neuraminidase. The envelope is pleomorphic and the diameter of the virions ranges from 120 to 200 nm. The viral genome codes for eight structural and two non-structural proteins. Based on analyses with monoclonal antibodies, RSV is divided into two antigenic subgroups (A and B). The major differences between these subgroups are attributed to the G surface glycoprotein which is responsible for virus attachment to host cells. There is no known difference in clinical properties between the two subgroups.

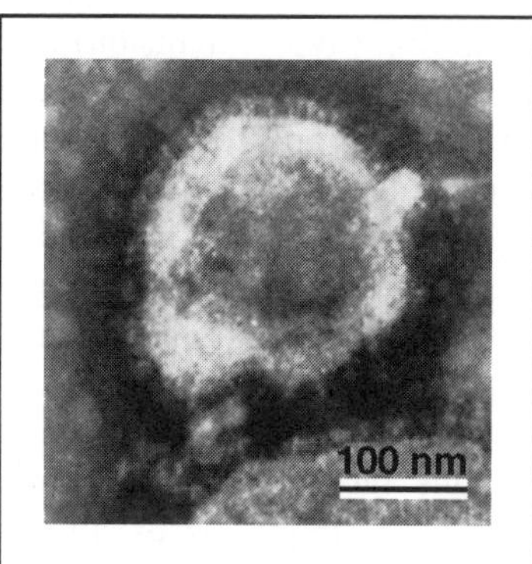

Figure 13.2 RESPIRATORY SYNCYTIAL VIRUS (Electron micrograph courtesy of E. Kjeldsberg)

EPIDEMIOLOGY

RSV has a worldwide distribution, and in temperate climates epidemics occur almost yearly during the colder months, whilst in tropical areas RSV outbreaks usually occur during the rainy season. In countries in temperate climates these epidemics are usually very regular in both size and timing. However, in some sparsely populated areas (e.g. Scandinavia) epidemics tend to alternate between greater and smaller outbreaks every second winter. RSV epidemics are characterized by distinct peaks usually reached 2–3 months after the first RSV cases are diagnosed. It has been claimed in some reports that during the peak of an RSV epidemic other outbreaks of respiratory virus (influenzavirus and parainfluenzavirus) infections seem to be relatively rare. Typical for a developing RSV epidemic is a sharp increase in number of infants and small children admitted to hospital with lower respiratory tract infection. The incidence of RSV infection severe enough to require hospitalization has been estimated to range from 1 to 3% of infants born each year. On the other hand, serological surveys have revealed that approximately half the infants living through a single RSV epidemic become infected and almost all children have been infected after living through two RSV epidemics. Thus, severe lower respiratory tract involvement is rather the exception, even among infants and small children. Among older children and adults reinfections are fairly common. These reinfections, which clinically contribute to the common cold syndrome, probably represent the major RSV reservoir, whilst infants and small children with severe lower respiratory tract involvement serve as the visible parameter of RSV activity within the community.

THERAPY AND PROPHYLAXIS

Good supportive care is of great importance in the treatment of patients with RSV-induced lower respiratory tract involvement. Since most hospitalized children are hypoxaemic, humidified oxygen is beneficial. In severe cases the use of a respirator may be necessary. Many infants are moderately dehydrated, and intravenous fluid replacement should be considered. Ribavirin given as an aerosol has been shown to have a beneficial effect, both on virus shedding and on clinical illness. Trials with monthly intravenous infusion of RSV **immunoglobulin** given to infants at high risk for severe RSV infection have indicated a prophylactic effect. The use of **vaccines**, either inactivated or attenuated formulations, has hitherto given disappointing results. As nosocomial RSV infections are common, hospitalized infants and children with suspected or proven RSV infection should be kept in isolation. Strict general hygienic measures should be instigated as the hospital staff may transmit RSV.

LABORATORY DIAGNOSIS

RSV antigen can be detected in exfoliated nasopharyngeal cells by immunological methods (IF, ELISA). Samples are collected with a tube connected to the outlet of a mucus collector. It is crucial to collect a sufficient amount of material. If the transportation time to the diagnostic laboratory is more than 3 to 5 hours, the samples should be processed at the clinical department, either by separating cells and mucus by a washing procedure or by making smears on slides of the untreated aspirated material. After drying the preparation can be sent to the diagnostic laboratory by ordinary mail. With the ELISA technique, the time factor is less important. If the sample is collected during the first week of illness, up to 90% of the actual RSV infections may be diagnosed with these methods. Several commercial kits for rapid bedside diagnosis are now available. The sensitivity and specificity of these kits are not fully evaluated, particularly when used at busy clinics. RSV infection can also be diagnosed by conventional virus isolation in tissue culture. However, viral infectivity is readily lost during transportation, and virus cultivation and identification are cumbersome and time-consuming (1–3 weeks). Significant titre rise can be detected in paired sera by serological methods, but among infants and small children serological tests for RSV are rather insensitive. High titres should be interpreted with caution.

VACCINATION MAKES ALL THE DIFFERENCE

14. MEASLES VIRUS

Lat. *morbilli*; Ger. *Masern*; Fr. *rougeole*.

N. A. Halsey

Measles is a highly contagious, serious disease affecting children, adolescents and occasionally adults.

TRANSMISSION/INCUBATION PERIOD/CLINICAL FEATURES

Measles virus is transmitted from respiratory secretions by direct contact, droplets or airborne transmission with inoculation onto mucous membranes. The incubation period is 10 (8–15) days.

SYMPTOMS AND SIGNS

Systemic:	Fever and Malaise
Local:	Rash, Cough, Coryza, Conjunctivitis, Koplik's Spots
Others:	*See* Complications

In uncomplicated cases the clinical disease improves by the third day of rash and resolves by 7–10 days after onset of rash.

COMPLICATIONS

The most common complications affect the respiratory tract and include otitis media, laryngotracheobronchitis (croup), pneumonia and secondary bacterial pneumonia. Encephalitis occurs in 1 per 1000 infected children. Subacute sclerosing panencephalitis (SSPE) is a rare, fatal complication appearing after an interval of 6–8 years.

THERAPY AND PROPHYLAXIS

No specific antiviral therapy. Measles vaccine, either alone or combined with mumps and rubella, is an effective prophylactic measure. After exposure, human immunoglobulin given up to 3–5 days after exposure is effective.

LABORATORY DIAGNOSIS

Antigen-capture measles-specific IgM antibody assays are available and are highly specific and sensitive. Confirmation of the diagnosis by demonstration of a 4-fold rise in measles-specific IgG antibodies in acute and convalescent sera. Virus can be isolated from throat, conjunctiva and urine (not routinely used).

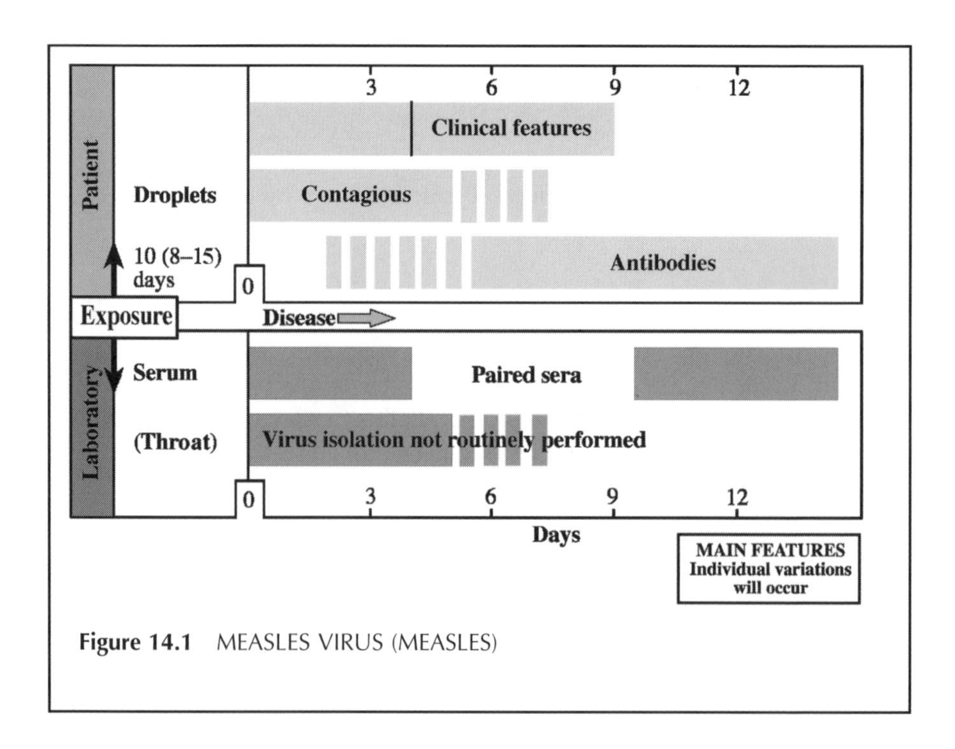

Figure 14.1 MEASLES VIRUS (MEASLES)

CLINICAL FEATURES

SYMPTOMS AND SIGNS

After an incubation period of approximately 10 (range 8–15) days, patients develop fever, cough, coryza and conjunctivitis which increase in severity for 2–4 days. On the day prior to the onset of rash, and for 1–2 days afterwards, Koplik's spots (2–3 mm diameter, bluish-white dots on an erythematous base and being pathognomonic of measles) appear on mucous membranes, especially the buccal mucosa in 70–80% of patients. After 3–4 days of illness a discrete maculopapular rash appears on the face and neck and spreads to the trunk, and temperature rises to 39–40°C. Lesions on the trunk and face may become confluent by the third day and then gradually fade. A fine, brawny desquamation appears 7–8 days after onset of the rash, but is often not noticed in children who are frequently bathed.

Differential diagnosis. Other viral infections have been mistaken for measles (parvovirus B19, rubella, enteroviruses, dengue, adenoviruses and Epstein–Barr). Other rash illnesses that have been confused with measles include eritherm multiforme, toxic shock syndrome, leptospirosis, Rocky Mountain spotted fever and scarlet fever, but these illnesses do not have the same clinical profile of measles with a prodromal respiratory infection and increasing fever for 2–4 days prior to onset of rash.

CLINICAL COURSE

In the absence of complications the clinical disease is usually improving by the third day of rash and resolves by 7–10 days. A mild modified course of measles may be noted 1 week after vaccination or when immuoglobulin is given in the incubation period. A more severe form of measles, dominated by high fever and haemorrhages from skin and mucosa, occurs rarely. Among persons suffering from protein malnutrition the lethality is high.

COMPLICATIONS

The most common complications affect the respiratory tract and include otitis media, laryngotracheobronchitis (croup) and interstitial pneumonia. Otitis media occurs in 5–25% of children less than 5 years of age. Pneumonia is seen in 5–10% of children under 5 years, and more frequently in adults. Diarrhoea occurs in approximately 10% of young children, croup less frequently, but the latter may be severe and life-threatening. A persistent diarrhoea with protein-losing enteropathy and (subsequent worsening of) ensuing malnutrition is seen in developing countries. Thrombocytopenic purpura has been reported on rare occasions and complications involving other organ systems have occasionally been seen. Encephalitis occurs in 1 per 1000 infected children. Subacute sclerosing panencephalitis (SSPE), a slowly progressive CNS infection, occurs in an average of 5–20 per million children who have had measles; the onset is delayed an average of 7 years after measles. The illness lasts from 1 to 3 years and inevitably leads to death. For measles the case-fatality rate averages 3 per 1000 children in the USA and between 3 and 30% of young children in developing countries.

THE VIRUS

Measles virus (Figure 14.2) is an enveloped RNA virus belonging to the genus *Morbillivirus* within the *Paramyxoviridae* family. The virus is related to canine distemper virus. Other members of genus have been shown to cause severe disease in mammals, e.g. seals and horses. Measles is a single-stranded virus of negative polarity surrounded by the nucleoprotein (NP), the phosphoprotein (P), the matrix protein (M) and a large (L) protein with polymerase function. The envelope contains the two viral glycoproteins: the haemagglutinin (H) and the fusion (F) protein. The H is responsible for the binding of measles virus to cells and the F protein for the uptake of virus into the cells. Antibodies to H correlate with protection against disease. Sequencing data of the genes coding for the H, NP and F proteins indicate a gradual change in strains isolated in various parts of the world. However, measles virus vaccines based on viruses isolated more than 30 years ago continue to confer immunity. SSPE isolates obtained by co-cultivation produce either a defective M protein or no M protein.

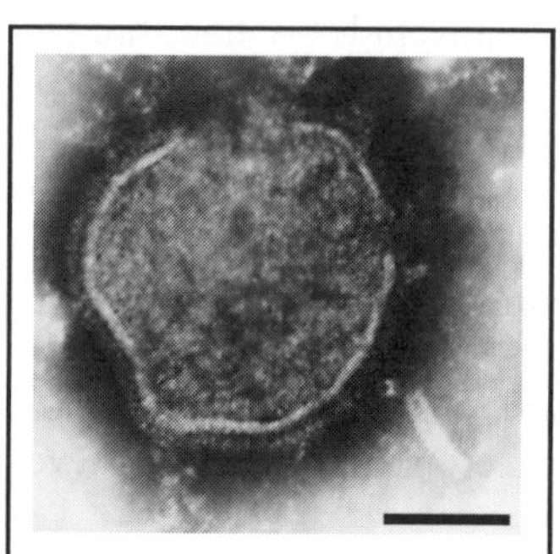

Figure 14.2 MEASLES VIRUS. Bar, 100 nm (Electron micrograph courtesy of D. Hockley)

EPIDEMIOLOGY

Persons are most infectious during the prodromal phase of illness when the virus is transmitted through aerosol droplets. Airborne transmission has been well documented. Most cases occur in the late winter and early spring, but low levels of transmission continue to occur year-round in most climates. In the tropics, most cases are seen during the dry season. When the measles virus is introduced into a non-immune population, 90–100% become infected and get clinical measles. Epidemics of measles occur every 2–4 years when 30–40% of children are susceptible. Immunity is lifelong. After the introduction of an effective vaccine, case reports have fallen by over 90%, widespread. Prior to widespread immunization, most cases in industrialized countries occurred in

children aged 4–6 years and in small children in developing countries. Intensive immunization coverage has resulted in a shift in age distribution, with relatively more cases among older children, teenagers and young adults. With widespread implementation of two-dose vaccination strategies, measles is now nearly eliminated in most European and Latin American countries. However, if a high vaccine coverage rate (>95%) is not upheld after measles has been eliminated in the community, accumulation of susceptibles will inevitably lead to recurrence of outbreaks or epidemics from imported cases as long as global eradication of measles is not yet achieved.

THERAPY AND PROPHYLAXIS

Supportive therapy only is indicated for most well nourished children. Antibiotic treatment is indicated for bacterial otitis media and pneumonia. Vitamin A (100,000 units for less than 12 months of age, 200,000 units for over 12 months of age) has resulted in 50% reduction of mortality in developing countries. In severe cases (including immunocompromised children), ribavirin has been administered systemically based on *in vitro* susceptibility testing, and some children have shown evidence of clinical response. Human **immunoglobulin** administered after exposure in a dose of 0.25 ml/kg modifies the disease in most children if given within 3–5 days after exposure. A dose of 0.5 ml/kg is recommended for immunocompromised children. An attenuated measles **vaccine** is available either as a single formulation or combined with attenuated mumps and rubella viruses (MMR). About 98% of children immunized at the age of 12–15 months develop an antibody response and vaccine efficacy is ⩾90% following a single dose. A second dose, given later in life, increases immunity to approximately 99%. 6–15 days after immunization 5–15% of children will develop a mild fever and/or rash. Neurological complications (e.g. encephalitis) occur in less than 1 in a million vaccinees. Most immunocompromised children should not receive the vaccine, but vaccination is recommended for HIV-positive children without severe immune suppression. Pregnancy is a contraindication to vaccination. Rare cases of immediate hypersensitivity have occurred to the gel stabilizer. The administration of passive antibody up to 12 weeks before vaccination will blunt or block the immune response dependent on the dose of immunoglobulin administered.

LABORATORY DIAGNOSIS

Antigen-capture measles-specific IgM antibody assays (ELISA) have been developed, and when used appropriately are highly sensitive and specific. The test becomes positive within 48 hours after rash and may remain positive for up to 30 days after the onset of illness. The standard method for confirming the diagnosis is demonstration of a 4-fold rise in IgG measles antibodies in acute and convalescent sera. Haemagglutination–inhibition tests or ELISA antibody assays are most practical, but plaque reduction neutralization tests are the most

sensitive and specific. The virus has been isolated from respiratory tract secretions and rarely from urine or circulating lymphocytes during the prodromal phase of illness or within a few days after the rash onset. Immunofluorescence staining of nasal or throat secretions or urine has been successful, but is not widely available. SSPE is confirmed based on characteristic EEG patterns and demonstration of measles antibody in the cerebrospinal fluid (CSF) with an increased CSF to serum measles antibody ratio, or by demonstration of virus in brain tissue.

Very high measles antibody titres aside from acute infection and SSPE are regularly seen in autoimmune chronic active hepatitis and occasionally in systemic lupus erythematosus.

A TIME TO AVOID INFECTION

15. RUBELLA VIRUS

German measles; Ger. *Röteln*; Fr. *rubéole*.

G. Haukenes

Rubella is a mild exanthematous, moderately contagious disease. When the disease is acquired by the mother during the first 4 months of pregnancy, the virus may infect the fetus and cause serious malformations.

TRANSMISSION/INCUBATION PERIOD/CLINICAL FEATURES

Transmission occurs by droplet inhalation. The incubation period is 2–3 weeks. The patient is infectious from 7 days before and up to 7 days after the appearance of the rash. A child with congenital rubella may excrete virus for up to 2 years.

SYMPTOMS AND SIGNS

Systemic:	Low-Grade Fever, Mild Malaise
Local:	Rash, Enlarged and Tender Lymph Nodes (Suboccipital, Postauricular, Posterior Cervical)
Other:	*See* Complications

The rash lasts for 3–4 days, the lymphadenopathy somewhat longer.

COMPLICATIONS

Postinfectious encephalitis, purpura, arthralgia, congenital rubella.

THERAPY AND PROPHYLAXIS

No specific antiviral therapy. The effect of specific immunoglobulin is uncertain. Vaccine provides about 95% protection.

LABORATORY DIAGNOSIS

Antibodies can be detected within a few days of appearance of rash. Diagnosis is based on seroconversion or rise in titre in paired sera taken at 1–2 week intervals, or by demonstration of specific IgM. Presence of antibodies indicates immunity to clinical reinfection. In congenital rubella, virus can be isolated from pharynx or urine for a period of up to 2 years, and specific IgM is usually present at birth.

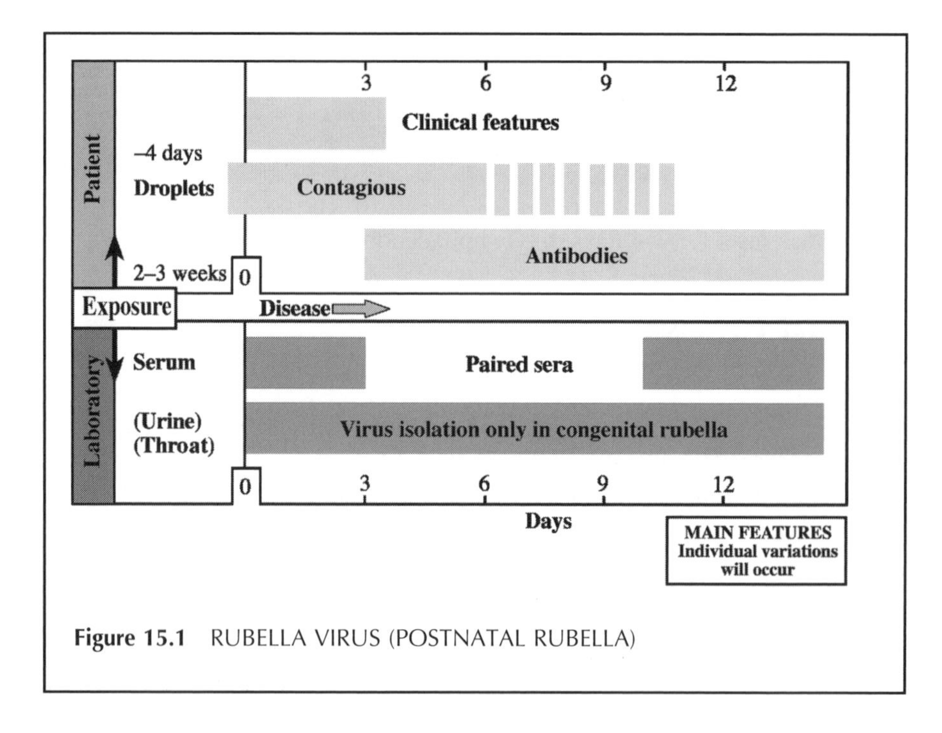

Figure 15.1 RUBELLA VIRUS (POSTNATAL RUBELLA)

CLINICAL FEATURES

POSTNATAL RUBELLA

SYMPTOMS AND SIGNS

Transmission occurs by droplet inhalation. The **incubation period** is 2–3 weeks, but may be prolonged after administration of specific immunoglobulin. Many cases run asymptomatically or atypically (without rash). Mild conjunctivitis and fever are sometimes observed as prodromes, especially among adults. The rash, which is pinpoint maculopapular, spreads from face to neck and over the trunk as pink, slightly raised macules, which may coalesce. There may be petechiae on the soft palate, conjunctivitis and mild catarrhal symptoms. Posterior cervical, postauricular and occipital lymph nodes are often enlarged and tender. The swelling may persist for 2–3 weeks. Fever is low grade or absent during the rash. Some patients have a more generalized lymphadenopathy and splenomegaly with slight fever during the rash, but the general condition is good. The white cell count is lowered with increased plasma cells and a relative increase in both small and large atypical lymphocytes (Türk cells).

Differential diagnosis. Other exanthematous infections: measles, scarlet fever, exanthema subitum, erythema infectiosum, infectious mononucleosis, echovirus and coxsackievirus infections. In particular, the possibility of erythema infectiosum (parvovirus B19) should be considered as both viruses may infect the fetus and may manifest with rash, suboccipital lymphadenopathy and arthralgia, and serological cross-reactivity in the IgM assay has been seen. There is no clinical picture that is pathognomonic for rubella. The diagnosis can be made with certainty only by serological tests.

CLINICAL COURSE

The rash rarely persists for more than 3 days. In about 40% of cases the infection is asymptomatic. Immunity to clinical and intrauterine reinfection is lifelong.

COMPLICATIONS

Postinfectious encephalitis is rare (1 in 5–10,000) and sequelae are seldom seen. Arthralgia is often seen in adult women and may persist for weeks and months. Some patients have transient thrombocytopenic purpura.

CONGENITAL RUBELLA

Rubella virus may infect the placenta and the fetus. About 30% of children born to mothers who have acquired rubella during the first trimester of pregnancy will have congenital defects at birth. The risk approaches 100% if rubella occurs in the first month, but falls to about 10–20% in the fourth month. Later in pregnancy, the risk of serious damage to the fetus is low. Rare cases of congenital rubella have been reported when the mother had rubella up to 1 month before becoming pregnant. In about 15% of cases infection of the fetus leads to spontaneous abortion. The classical triad of malformations in congenital rubella involves the eye (cataract, microphthalmia, glaucoma, chorioretinitis), the heart (most often patent ductus arteriosus) and the ear (unilateral or bilateral sensoryneural deafness). Psychomotor retardation is also commonly seen. Most affected infants have a low birth weight but normal body length. Purpura, hepatosplenomegaly, hepatitis, anaemia and pneumonia may be present.

Differential diagnosis. This includes other intrauterine infections such as toxoplasmosis, syphilis and CMV infection. The precise diagnosis is made by serological tests and demonstration of virus.

THE VIRUS

Rubella virus (Figure 15.2) is a positive-sense single-stranded RNA virus of the genus *Rubivirus* within the *Togaviridae* family. The virion is medium-sized with an outer lipoprotein membrane ('toga'). Three structural proteins have been identified, of which two are glycoproteins located in the envelope (E1 and E2), and the third is a non-glycosylated core protein (C). Only one antigenic type of the virus is known. The structural proteins are cleaved from a precursor protein translated for subgenomic 24S mRNA. E1 and E2 become N-glycosylated in the endoplasmatic reticulum. Virus matures by budding from the plasma membrane of intracytoplasmic membranes. The virus may be grown in a wide range of cell cultures, with or without cytopathic effect. The fetal damage seen in congenital rubella seems to be partly due to virus-induced retardation of cell division which, when it occurs in the early phase of organogenesis, may lead to serious malformations. In addition, some endothelial damage and cell necrosis are seen. Infection of the fetus before the immune response has been established results in persistence of virus throughout gestation and for months or years after birth.

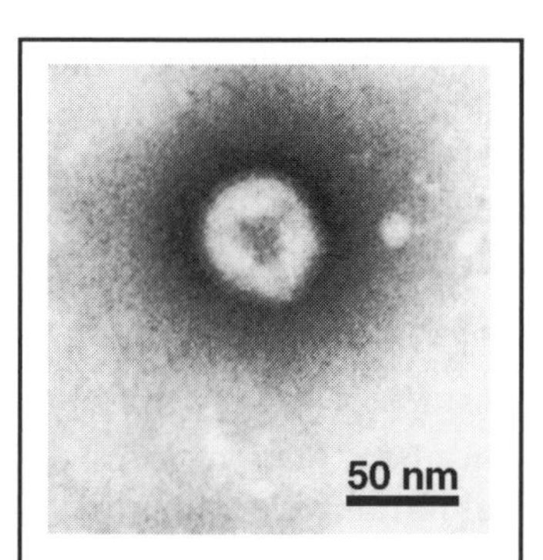

Figure 15.2 RUBELLA VIRUS. Bar, 50 nm (Electron micrograph courtesy of E. Kjeldsberg)

EPIDEMIOLOGY

Rubella is worldwide in distribution. In postnatal rubella the patient is considered to be infectious from up to 7 days before to 7 days after the onset of rash. Transmission occurs by droplet inhalation. Children with congenital rubella may shed large amounts of virus in the pharynx, faeces and urine for a long period (up to 2 years). Epidemics of rubella of 1 to 2 years' duration are seen at about 5 year intervals. During an epidemic there are about 10 times more cases than in the interepidemic period. In temperate climates most cases are seen in spring and early summer. The disease is most prevalent among children of 5–10 years of age. At puberty about two-thirds of all children are seropositive. Serological examination of pregnant women shows that 80–95% of them are seropositive. Susceptible pregnant women usually contract the disease from (their own) children. Teachers in kindergartens and primary schools are professional groups at special risk. Vaccination of girls before puberty and special risk groups aims to render most childbearing women immune. However, such immunization programmes lead to only moderate reduction in the amount of virus circulating in the community. Vaccination coverage will usually not exceed 90% and therefore some fertile women remain susceptible and at risk. Many countries have therefore implemented a routine immunization programme for all infants at the age of about 15 months, in some countries followed by revaccination of both sexes before puberty. As a result, very few cases of rubella are seen, and the cases of congenital rubella may be completely avoided within a few years in these countries.

THERAPY AND PROPHYLAXIS

There is no specific antiviral **therapy**. In congenital rubella the disease manifestations at birth may require hospital treatment. Some of the malformations are treated surgically. High-titre rubella **immunoglobulin** seems to have a certain protective effect when given early after exposure. Administration of immunoglobulin may lead to asymptomatic infection and a prolonged incubation period. This has to be taken into account during the serological follow-up that should be carried out in these cases. The **vaccine** consists of live attenuated virus cultivated in a human fibroblast diploid cell line. Vaccination provides long-lasting immunity in about 95% of those vaccinated. Side-effects such as mild rubella symptoms are seen in 20%, and arthralgia occurs especially in women. There are no contraindications to vaccination other than those against the use of live virus vaccines in general: immune deficiency, active infection and pregnancy. Fetal infection has been observed in a very few cases after inadvertent vaccination of seronegative pregnant women, but no teratogenic effects of the vaccine have been seen. Nevertheless, women should not become pregnant in the 3 months after vaccination. The vaccine is dispensed as a monovalent vaccine or as a trivalent vaccine together with measles and mumps virus (MMR).

LABORATORY DIAGNOSIS

Postnatally two diagnostic problems may arise, acute infection and immunity. The clinician should therefore always state the clinical problem, so that proper diagnostic methods can be chosen. In **acute infection**, serum sample(s) are examined for the presence of specific IgM using an indirect or IgM capture ELISA. If an indirect ELISA is used, measures should be taken to avoid interference by rheumatoid factor (absorption of IgG). The IgM (anti-μ) capture method is generally considered to be the most specific one. IgM antibodies can be demonstrated a few days after appearance of rash and persist for about 3 months. Also IgG (anti-E1) antibodies appear early and usually persist for life. Occasionally a positive IgM test is seen in infectious mononucleosis (if the capture assay is used, preabsorption of heterophile antibodies by sheep red cells can be attempted), in autoimmune chronic active hepatitis type 1 and in parvovirus B19 infection. A positive IgM test is also seen in some cases of reinfection. In early pregnancy it is of crucial importance to exclude causes of a positive IgM test other than a primary rubella infection. An IgG seroconversion proves a primary infection. Since antibodies to E2 appear several weeks after the rash, anti-E2 seroconversion may also be looked for (by passive haemagglutination based on E2-coated erythrocytes or by Western blot). Several IgM ELISAs should also be tried. **Immunity** may be screened by several methods (ELISA, latex agglutination, HI) which specifically detect at least 15 international units (I.U.) of anti-E1 antibodies. In most cases of autoimmune chronic active hepatitis very high levels (up to 90,000 I.U.) of anti-E1 IgG rubella antibodies (and/or measles antibodies) are seen, which may be of differential diagnostic value versus primary biliary cirrhosis where low titres are found.

In **congenital rubella** virus may be isolated from urine and pharynx using a rabbit kidney (RK13) or cornea (SIRC) cell line. When cytopathic changes appear, the virus can be identified using the immunofluorescence method. The finding of IgM antibodies and persistence or rise of IgG antibodies during the first year of life is also diagnostic of congenital rubella.

A DAY OUT FOR THE ADENOVIRUS FAMILY

16. ADENOVIRUSES

The name originates from adenoids, the nasopharyngeal lymphoid tissue from which virus was first isolated.

I. Ørstavik and D. Wiger

Adenoviruses cause infection of the respiratory tract, the eye and the intestine.

TRANSMISSION/INCUBATION PERIOD/CLINICAL FEATURES

The infection is transmitted by droplets or by contact with virus-contaminated objects (e.g. hands, towels, medical equipment). The incubation period varies from 5 to 10 days. The patient is usually contagious as long as the symptoms last.

SYMPTOMS AND SIGNS

Systemic:	Fever
Respiratory:	Nasopharyngitis, Occasional Pneumonia (Especially in Infants)
Ocular:	Keratoconjunctivitis, Conjunctivitis
Gastrointestinal:	Diarrhoea, Vomiting

Respiratory symptoms last for about 1 week, ocular symptoms for 2–8 weeks and gastrointestinal symptoms for up to 10 days.

COMPLICATIONS

Rare fatal cases of pneumonia and disseminated adenovirus infection have been reported. Permanent opacity of cornea is rare.

THERAPY AND PROPHYLAXIS

There is no specific treatment. A vaccine has been used for immunization of military recruits.

LABORATORY DIAGNOSIS

Adenoviruses can be detected by isolation in nasopharyngeal and conjunctival secretions and throat washings and are seen by electron microscopy of faeces during the acute phase of the disease. A rise in CF antibodies to a common antigen is usually seen in respiratory infections. Viral antigens may also be detected in clinical specimens using various immunological techniques and PCR.

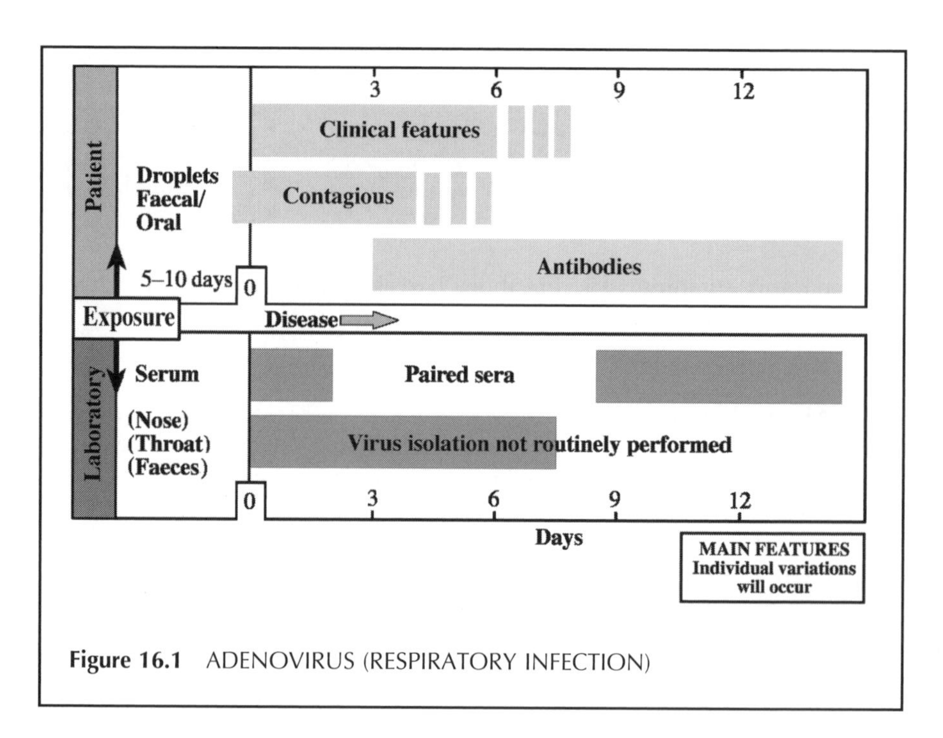

Figure 16.1 ADENOVIRUS (RESPIRATORY INFECTION)

CLINICAL FEATURES

SYMPTOMS AND SIGNS

The incubation period is usually between 5 and 10 days and varies according to the infective dose and the clinical manifestation. The patient is contagious as long as the symptoms last, but the virus may persist in tonsils and adenoids.

Respiratory disease. In acute upper respiratory tract infections symptoms are fever, nasal congestion and sore throat with a cough. This type of infection occurs most often in preschool children, although about 50% of adenovirus infections in this age group are asymptomatic. Serious bronchopneumonia may occasionally occur, especially in infants.

Acute respiratory disease (ARD) occurs in epidemics primarily among military recruits about 3–4 weeks after they have entered service. Sore throat, cough and fever last about 1 week. Lung infiltrations may be found, but overt pneumonia is rare.

Pharyngoconjunctival fever often occurs in children and presents with pharyngitis, conjunctivitis and fever, often accompanied by rhinitis, otitis media and/or diarrhoea. Exudate may be present on the tonsils and posterior pharyngeal wall. Cervical lymphadenopathy is common.

Ocular disease (without respiratory symptoms). A follicular conjunctivitis is seen with oedema of periorbital tissue, redness of the conjunctivae and serous exudation. Recovery is usually complete.

Epidemic keratoconjunctivitis (EKC) starts as an acute unilateral conjunctivitis that often eventually involves both eyes. Preauricular lymphadenopathy with some tenderness is observed. Subepithelial opacities in the cornea may last from 2 to 8 weeks, occasionally years, or permanently.

Gastrointestinal disease. Infantile diarrhoea gives moderate diarrhoea, vomiting and fever.

Certain adenoviruses have been associated with haemorrhagic cystitis, intussusception in infancy and a whooping cough-like disease.

Persistent and sometimes severe infection with certain adenoviruses has been observed in AIDS patients and in other immunocompromised individuals. Severe adenovirus infections have been observed in children with severe combined immunodeficiency (SCID).

In pharyngitis and ARD, **differential diagnoses** are infections with *Streptococcus pyogenes*, mycoplasma, chlamydia (psittacosis), coxiella (Q-fever) and various respiratory viruses. An infection with *S. pyogenes* is usually more aggressive and involves primarily the tonsillar tissue. Pharyngoconjunctival fever often occurs in epidemic form and is usually easier to diagnose. In follicular conjunctivitis bacteria (*Chlamydia trachomatis*) should be considered. Epidemic haemorrhagic keratojunctivitis is usually caused by enterovirus 70, while unilateral keratitis without conjunctivitis suggests herpes simplex virus.

Laboratory confirmation is necessary in order to establish a definite aetiological diagnosis.

CLINICAL COURSE

Gastroenteritis lasts up to 10 days and uncomplicated respiratory infections about 1 week. The eye infection may persist for 2–8 weeks or longer.

COMPLICATIONS

Secondary bacterial infections (otitis media and sinusitis) are occasionally seen in cases of respiratory tract infections. Some cases of lung fibrosis and bronchiectasis have been reported, especially in connection with serotype 7 infections. Keratitis may in rare cases lead to permanent opacities of the cornea with impairment of vision.

THE VIRUS

Human adenoviruses are medium-sized non-enveloped viruses with a diameter of about 80 nm (Figure 16.2) and include 47 different species (serotypes). The virus capsid is an icosahedral shell made up of 252 capsomers, 240 hexon capsomers and 12 penton capsomers at the vertices of the icosahedron. Pentons have antenna-like projections (fibres) which vary in length depending on the subgenus of virus. The virus capsid encloses a double-stranded DNA protein complex, and the virus contains at least 10 different polypeptides. The human adenovirus genus consists of 47 or more serotypes which are divided into six subgenera (A–F), based on DNA homology. The viruses in the different subgenera often share biological, clinical and epidemiological characteristics. The type-specific antigens are located on the outside of the hexons and the fibres. Antibodies to these antigens seem to be protective and probably last throughout life. The virus serotype is based on virus neutralization by antibodies against these antigens. Human adenovirus common group antigens are also found on the hexons, but are probably located internally. These cross-reacting antigens, believed not to stimulate protective antibodies, are used in CFT and ELISA. Adenoviruses are somewhat resistant to many physical and chemical agents. Because of this, virus is easily spread by contact with virus-contaminated equipment and water.

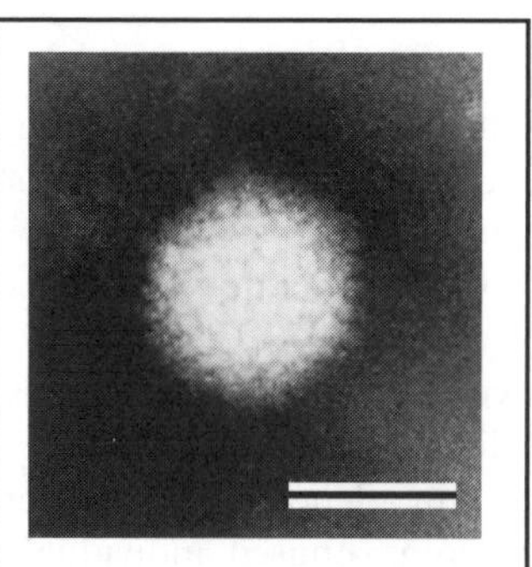

Figure 16.2 ADENOVIRUS. Bar, 100 nm (Electron micrograph courtesy of E. Kjeldsberg)

EPIDEMIOLOGY

Virus is spread by the faecal–oral route, especially in families. The respiratory route (aerosols) has been shown to be important in the spread of respiratory infections among military recruits. The eye is also an important portal of entry for certain adenoviruses. Adenoviruses 1, 2, 5 and 6 are endemic among preschool children and are associated with respiratory disease in this age group. These viruses are known to give prolonged infections of the lymphatic tissue in the throat and intestine, with continual or intermittent virus shedding. Because of this, isolation of these viruses will not always be diagnostic of a current infection. Adenoviruses 3, 4, 7, 14 and 21 have caused epidemic outbreaks of respiratory disease among military recruits as well as sporadic infections in the civilian population. Conjunctivitis has been reported to be caused by serotypes 1, 2, 3, 4, 6, 9, 10, 15, 16, 17, 20, 22 and 29. Adenoviruses 8, 19 and 37 have most often been associated with epidemic keratoconjunctivitis. Contaminated ophthalmological instruments (especially tonometers) have been implicated in the spread of this disease. Intrafamilial spread of these viruses is also important. Adenoviruses 40 and 41 cause diarrhoea and fever in children. Cases with adenovirus 41 tend to run a protracted course. Infections with these faecal adenoviruses are less often associated with respiratory symptoms than other types of adenoviruses. Many adenovirus species are isolated only occasionally, and little is known about their medical importance. Outbreaks of respiratory disease due to adenoviruses among military recruits occur mostly during the winter months in temperate areas, while adenovirus gastroenteritis is most frequently observed during the late autumn and winter. Otherwise, there seems to be very little seasonal variation in the occurrence of adenoviruses.

THERAPY AND PROPHYLAXIS

Specific **antiviral chemotherapy** or **immune therapy** is not available, and all treatment is symptomatic. A vaccine against certain types of adenovirus has been given to military recruits in some countries. Because nosocomial spread of adenovirus respiratory infections and gastroenteritis occurs quite often, isolation of patients in hospital is important. The spread of epidemic keratoconjunctivitis can be prevented by strict hygiene during eye examinations and by sterilization of equipment. The spread of virus within households may be controlled by hygienic measures such as careful handwashing.

LABORATORY DIAGNOSIS

Samples for virus isolation should be taken as early as possible in the course of the disease. Virus can be isolated in cell cultures from throat swabs, nasopharyngeal washings, rectal swabs and faeces from patients with respiratory infections. Virus isolation from the eye is usually made from swabs of the lower conjunctival sac or from conjunctival scrapings. Isolation in cell culture takes from a few days to 2–3 weeks. The adenoviruses that cause gastroenteritis often grow very poorly or not at all in conventional cell cultures. These viruses may be detected directly in faeces by electron microscopy or by immunological techniques such as latex agglutination or ELISA. Virus may also be detected by PCR. Rapid identification of adenovirus antigen in nasopharyngeal secretions, conjunctival scrapings or cell sediment from urine is also possible by immunofluorescence and ELISA. The CF antibody titre of infected patients (adults) shows a 4-fold or greater rise in paired sera. CF antibodies do not persist for long periods unless there are reinfections. More sensitive techniques, such as ELISA, may be necessary to detect the immune response to adenoviruses in small children. Infections of the eye alone may not give antibody titre rises detectable by CF tests.

RAPID SPREAD—DIARRHOEA ON WHEELS

17. ROTAVIRUSES

Lat. *rota* = wheel, reflecting its electron microscopic appearance.

I. Ørstavik and E. Kjeldsberg

Rotaviruses cause acute, often febrile gastroenteritis. The disease occurs most often in children, but may affect other age groups. In temperate climates most cases occur during winter/spring.

TRANSMISSION/INCUBATION PERIOD/CLINICAL FEATURES

The infection is spread by the faecal–oral route. Viral excretion in faeces may last for 4–7 days after onset of illness. The incubation period is 2–3 days.

SYMPTOMS AND SIGNS

Systemic:	Fever (Often High-Grade)
Local:	Diarrhoea, Vomiting

The illness usually lasts for 4–7 days. Complete recovery is the rule.

COMPLICATIONS

Febrile convulsions may occur in small children.

THERAPY AND PROPHYLAXIS

No specific therapy is available, but it is often necessary to correct the patients' fluid–electrolyte balance. Hospitalization may be necessary. Strict hygienic measures are needed to avoid further spread.

LABORATORY DIAGNOSIS

Rotavirus infection is diagnosed by demonstrating viral antigens by ELISA or latex agglutination, or viral particles by electron microscopy of faecal samples.

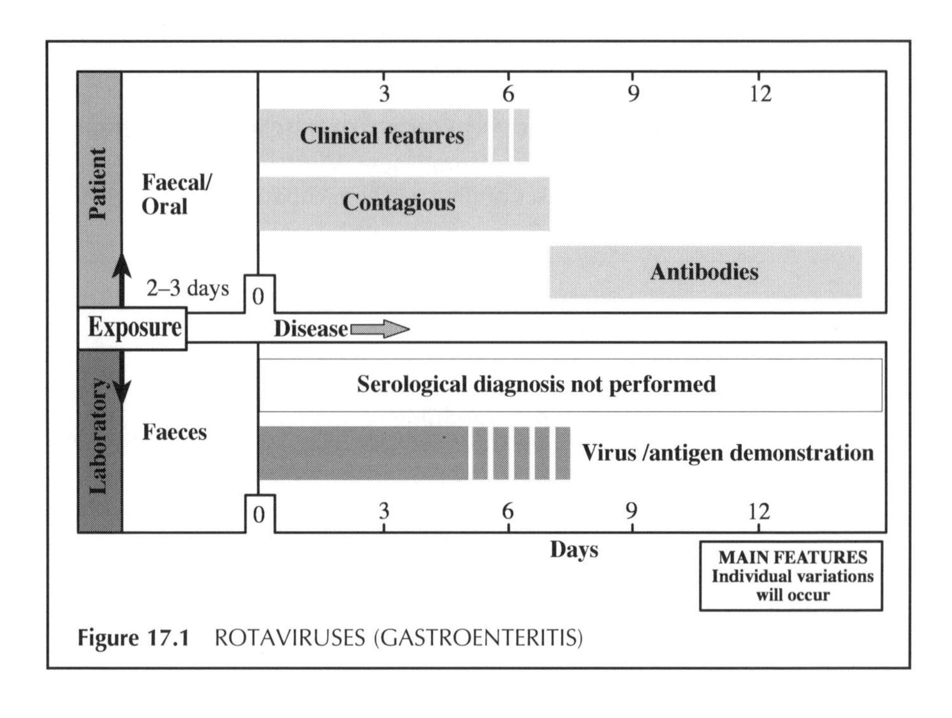

Figure 17.1 ROTAVIRUSES (GASTROENTERITIS)

CLINICAL FEATURES

SYMPTOMS AND SIGNS

After an incubation period of 2–3 days the illness usually begins with nausea and vomiting, followed by diarrhoea within 24 hours. Vomiting lasts for a short period, and is absent in some cases. In severe cases the stools are frequent, watery, voluminous and foul-smelling. Fever is commonly present and may be high-grade. In children febrile convulsions may occur. Dehydration is most often isotonic, but may be hypotonic, particularly when the patient has been given fluid deficient in electrolytes (e.g. juice). Visible blood in the stools is almost never observed. The white cell count is normal with relative lymphocytosis and transient neutropenia. An important factor in the pathogenesis of a rotavirus infection is lowered disaccharidase function in the gut. Disaccharides (lactose, sucrose) remain unabsorbed in the gut lumen and cause dehydration by osmotic pressure. Many patients with rotavirus gastroenteritis also have symptoms of upper respiratory tract infection, but it is not clear whether this is caused by rotavirus.

Differential diagnosis. The triad of vomiting, diarrhoea and fever is a typical finding in rotavirus gastroenteritis, but may be seen in gastroenteritis caused by other infectious agents. In shigellosis and salmonellosis there is usually leukocytosis/granulocytosis in peripheral blood, suggestive of a bacterial aetiology, and there may be visible blood in the stools. In gastroenteritis caused by *Escherichia coli* vomiting and fever are infrequent. Gastroenteritis caused by other viruses (adenovirus, astrovirus, calicivirus, Norwalk agent, etc.) is usually less severe than rotavirus gastroenteritis. However, as the symptoms and severity of rotavirus gastroenteritis may vary considerably, virological examination is necessary in order to obtain an aetiological diagnosis.

CLINICAL COURSE

The illness usually lasts 4–7 days, but cases have been described with protracted diarrhoea and virus excretion. The upper part of the small bowel is the main site for rotavirus infection, and damage to the distal parts of the intestinal villi with decreasing enzyme secretion probably causes a malabsorption resulting in diarrhoea. Serological studies have shown that many rotavirus infections are asymptomatic. Without dietary measures the course is often protracted, with recurrent bouts of diarrhoea. Severe cases with dehydration, hypotonia and shock may have a fatal outcome if the patient is not adequately rehydrated. Otherwise recovery is complete.

COMPLICATIONS

The only complication is febrile convulsions in children.

THE VIRUS

Rotaviruses are members of the genus *Rotavirus* within the family *Reoviridae*. The name of the virus is derived from the Latin word *rota*, meaning wheel, reflecting its shape as seen in the electron microscope. The intact virion of about 102 nm in diameter displays spikes from a smooth surface of about 79 nm (Figure 17.2(a)). When stools of patients are examined by electron microscopy, however, incomplete virus particles predominate, often without the smooth outer layer (Figure 17.2(b)). The virions contain a central double-stranded RNA consisting of 11 segments. The RNA segments code for a structural or non-structural protein. The antigenic structure and classification of rotaviruses are complex. At present six groups (A–F) are considered. Most rotaviruses found in humans belong to group A, but infections with groups B and C have also been reported. Group- and subgroup-specific antigens are located on the inner capsid layer, whereas type-specific antigens are found on the outer layer. Among group A rotaviruses, two subgroups and 12 serotypes have been described, of which six serotypes have been found in man. A typing system based on two structural proteins (G + P) is used, and of at least 14 group AG types, 10 have been detected in humans. The segmented genome must account for the complex antigenic properties of the rotaviruses, allowing reassortation of the 11 segments of RNA in cells being doubly infected with different strains. This mechanism is also described for influenzavirus. Rotaviruses are inactivated by heating to 100°C or treatment with acid ($pH < 3$), glutaraldehyde (3%) or alcohol (> 70%), while iodoform is less effective.

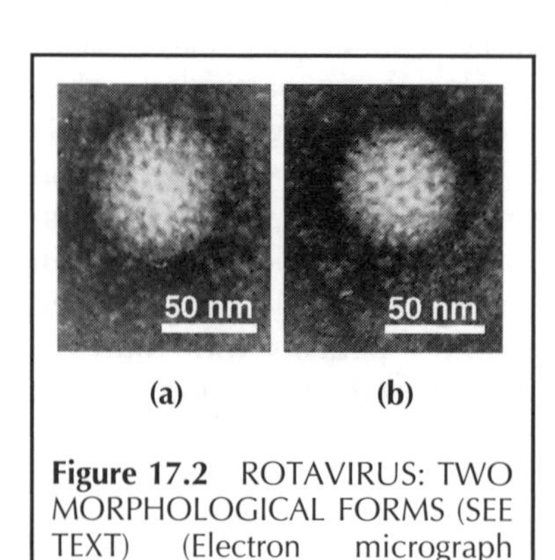

Figure 17.2 ROTAVIRUS: TWO MORPHOLOGICAL FORMS (SEE TEXT) (Electron micrograph courtesy of E. Kjeldsberg)

EPIDEMIOLOGY

Rotavirus infections occur all over the world. The majority of children have antibodies by the age of 5. Most clinical cases occur in children between 6 months and 3 years of age. Rotavirus is endemic, but in temperate zones most cases occur in winter/spring. In tropical and subtropical areas reports suggest that most cases occur during the rainy season. Up to 50% of hospitalized children with gastroenteritis are infected with rotavirus. The virus is spread by the faecal–oral route, but the rapid spread of the infection within institutions with high hygienic standards suggests that the virus is also spread by droplets from the throat, although this is not proven. Most newborn infants below

6 months of age are probably protected by circulating maternal antibodies and/or by factors in breast milk. One individual may become infected with different rotavirus serotypes and also be reinfected with the same serotype. Thus, although type-specific neutralizing antibodies are formed following infection, the protective effect may not be long-lasting. Extensive outbreaks of rotavirus gastroenteritis have been observed in hospitals and other institutions, and in a family setting there is often more than one case of illness in the same household. Most cases of human rotavirus gastroenteritis are caused by group A rotaviruses. Group B viruses have caused large epidemics in China. Rotavirus gastroenteritis causes a large number of deaths among children in developing countries, whereas fatal cases are very rare in developed countries.

THERAPY AND PROPHYLAXIS

There is no specific therapy. Symptomatic treatment appropriate for cases of acute gastroenteritis should be given with particular emphasis on oral rehydration with glucose–electrolyte solution. Milk may maintain the diarrhoea (see above), and should be omitted from the diet during the whole course of the illness. Admission to hospital should be considered if there are severe symptoms or if the patient is very young. Reports on **vaccination** as well as on the use of **immunoglobulin** prophylaxis have been published, but such measures are not generally available. An oral rotavirus vaccine for infants was introduced in the USA in 1998, but was withdrawn the following year because it was associated with an increased risk of intussusception. When a patient is treated at home, those caring for the patient should wash their hands frequently and wash carefully all items that may be contaminated with faeces and which cannot be boiled. In hospitals, isolation of the patient and strict hygienic measures are necessary to avoid cross-infection.

LABORATORY DIAGNOSIS

A specimen of faeces (at least 1 cm^3) should be obtained during the acute phase of the illness, and sent to the laboratory without any additives or refrigeration. Several rapid methods are available to detect virus or viral antigens in stools (electron microscopy, ELISA, latex agglutination, etc.), so the laboratory may often be able to report the results on the same day. Human rotaviruses can be cultivated in special cell cultures only with difficulty. Serological tests are not used for routine diagnosis of rotavirus infections.

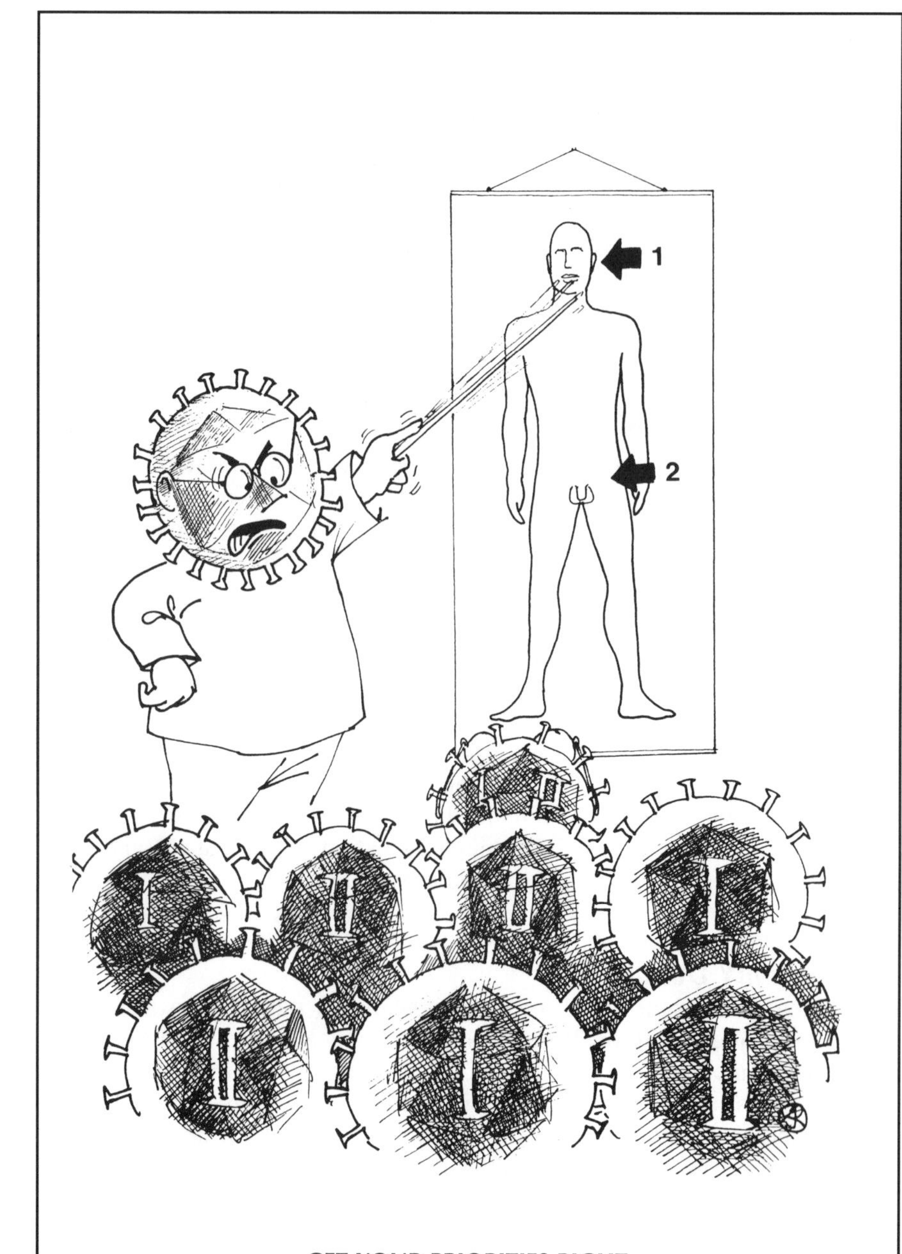

GET YOUR PRIORITIES RIGHT

18. HERPES SIMPLEX VIRUS (HSV1 AND HSV2)

From Greek *herpein* = creeping.

E. Tjøtta and G. Hoddevik

Herpes simplex virus infections occur worldwide. Lifelong latency is established after the primary infection. Recurrences as herpes labialis (usually HSV1) and herpes genitalis (usually HSV2) occur frequently.

TRANSMISSION/INCUBATION PERIOD/CLINICAL FEATURES

Infection by droplets or contact is most common. The incubation period is 2–12 days.

SYMPTOMS AND SIGNS

Systemic:	*Primary*:	Fever, Malaise. In Newborn Infants: Generalized Infection with a Septicaemia-like Picture
Local:	*Primary*:	Vesicles in the Mouth (Gingivostomatitis) or in the Genital Region, Lymphadenopathy
	Recurrent:	Vesicles on the Skin Near the Mouth ('COLD SORES') or in the Genital Region

Primary HSV1 infection is often clinically inapparent, but gingivostomatitis of about 2 weeks' duration occurs. Recurrent genital herpes does not last as long as the primary infection.

COMPLICATIONS

Keratitis may cause visual field impairment and blindness. Encephalitis is a rare but serious complication. Disseminated HSV infection is a serious complication in immunocompromised individuals. Eczema herpeticum (Kaposi's varicelliform eruption) occurs in children with a history of eczema. A herpetic whitlow is seen after accidental inoculation.

THERAPY AND PROPHYLAXIS

No treatment is used in recurrent, uncomplicated herpes. Complications or serious forms are treated with antiviral drugs. Vaccines are not yet available for general use. Testing of candidates showed limited promise.

LABORATORY DIAGNOSIS

Mucocutaneous herpes can be diagnosed by virus isolation or by direct demonstration of antigen. Antibody titre rise or seroconversion can be expected only in primary infection. Differentiation of type 1 and 2 is made by means of monoclonal antibodies or PCR. For diagnosis of encephalitis, see p. 133.

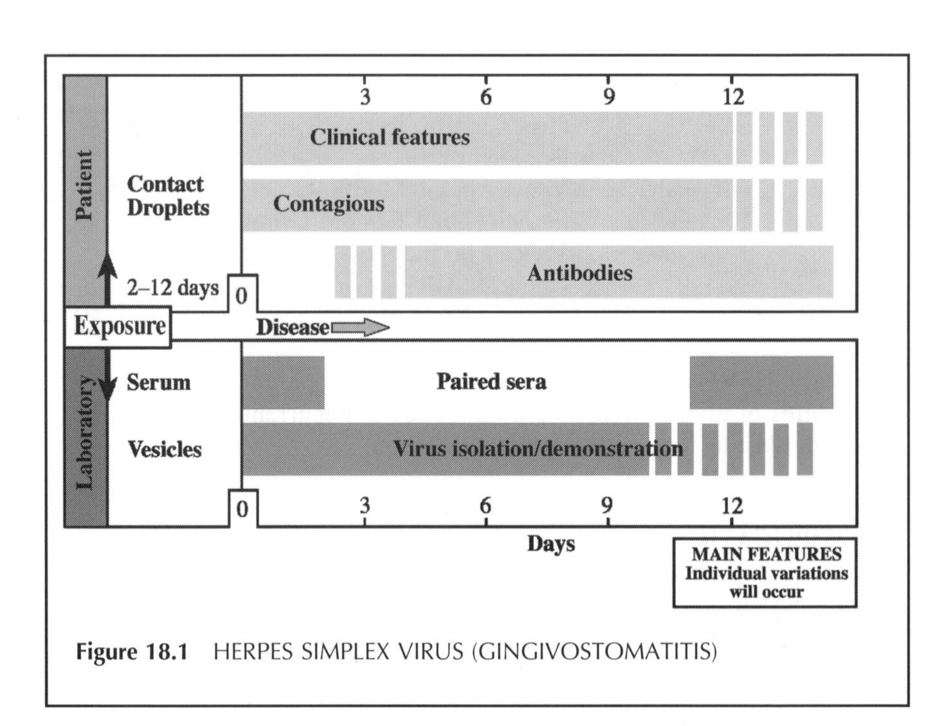

Figure 18.1 HERPES SIMPLEX VIRUS (GINGIVOSTOMATITIS)

CLINICAL FEATURES

SYMPTOMS AND SIGNS

HSV1 infections usually affect the oral cavity, lips or face, and HSV2 infections the genital region, but there is considerable cross-over, especially of HSV1 to the genital region.

In **Herpes 1**, labial herpes ('cold sore') is the most common manifestation. The primary infection is symptomless in most cases, but may present as fever, enlarged submandibular lymph nodes, sore throat, gingivostomatitis with ulcers or vesicles, oedema, with associated anorexia, pain and malaise. This condition usually lasts for 10–21 days, and may be accompanied by inability to eat or drink. Dehydration may be a problem, especially in small children. Symptomatic primary infection is most common in children of 1–5 years of age, with an incubation time of 2–12 days, mean about 4 days. There may be a prodrome of burning, itching or tingling pain for some hours followed by groups of vesicles usually on the external borders of the lips. Lesions may also be found in skin surrounding the lips; chin, cheeks or nose. Within a few days the vesicles progress to pustules or ulcers with brownish-yellow crusts. Pain is most severe in the beginning and resolves during the next 4–5 days.

The **Herpes 2** infection usually affects the genital regions. The primary genital infection may be severe, with illness usually lasting up to about 3 weeks (sometimes longer) with a shedding period of virus usually terminating shortly before or at the time of healing. The lesions are vesicles or ulcers localized to the cervix, vagina, vulva or perineum of the female, or the penis in the male. The lesions are painful, and may be associated with inguinal lymphadenopathy and dysuria. Systemic complaints, including fever and malaise, usually occur. Complicating extragenital affections, including aseptic meningitis, have been observed in about 10–20% of cases. Paraesthesia or dysesthesia may occur after the genital affection. Especially in women the severity of the primary infection may be associated with a high number of complications and frequent recurrences. Previous HSV1 infection reduces the severity and duration of primary HSV2 infection. The recurrent genital affection is usually milder, with fewer vesicles or ulcers and with a duration of 7–10 days. The recurrent lesions seldom last more than 10 days, and the shedding of virus may terminate sooner. Sometimes, virus excretion can occur between active periods.

Differential diagnosis. The gingivostomatitis can be mistaken for Vincent's angina. In herpangina (coxsackie A virus) the lesions are fewer and smaller, localized to the pharynx and back of the mouth, and there is no gingivitis. Recurrent aphthous stomatitis is not caused by HSV. Skin lesions should be distinguished from impetigo and syphilis. Zoster may occasionally be a diagnostic problem. Eczema herpeticum (see Complications) may resemble varicella-zoster clinically. The encephalitis has to be distinguished from a brain abscess, and this can be done with computer-assisted tomography. Encephalitis caused by VZV is clinically very similar to herpes encephalitis. Similarly, a postinfectious encephalitis (measles, rubella, varicella) should be considered. Generalized disease in the newborn can also be caused by coxsackie B virus, and it is essential to exclude a bacterial sepsis.

COMPLICATIONS

Infection in the newborn may be acquired *in utero*, at or just after birth. The newborn has low resistance to this infection, and usually develops severe disease. The mortality rate of untreated disease is about 50%. Babies with neonatal herpes infection may develop:

- A disseminated generalized form with many affected organs, including the central nervous system (CNS).
- Encephalitis with or without herpetic lesions of the skin.
- Herpetic lesions localized to skin, mouth and eyes.

The generalized form is especially serious, and is often combined with intravascular coagulopathy, hepatic and adrenal necrosis, pneumonitis and/or encephalitis followed by permanent neurological sequelae if the patient survives. The congenital infection may induce malformations such as microcephaly or microphthalmia, or other symptoms such as jaundice, hepatosplenomegaly, bleeding diathesis, seizures, irritability, chorioretinitis and herpetic vesicles of the skin.

HSV encephalitis. In the USA HSV1 is considered the most common viral strain of fatal encephalitis. The annual incidence of HSV encephalitis is estimated to be 1 in 200,000. The lesion is usually a local process in the brain, consisting of haemorrhagic necrosis and oedema, mimicking a brain tumour. The localization is usually one of the temporal lobes. At later stages, however, the expansion retracts leaving scar tissue and midline structures deviating to the affecting side.

In **immunocompromised patients** the HSV infection may be severe, especially if the cellular immunity is reduced. This is true both in patients with a disease affecting the immune system, e.g. AIDS, and in patients under immunosuppressive treatment. Especially bone marrow, renal and cardiac transplant recipients are at risk for severe herpes infections. The lesions may be progressive, and involve unusual sites such as the respiratory tract, oesophagus, liver and intestinal mucosa, or occur as a disseminated infection in severely immunocompromised patients. The severity of the disease is directly related to the degree of immunosuppression, and will also last longer than usual, about 6 weeks. Malnourishment, especially in children, seems to aggravate symptoms. Even immunocompromised patients may discharge virus asymptomatically.

The initial infection with herpesvirus may be located in the eye, with severe **keratoconjunctivitis** as a result. Recurrent infections of the eye may appear as ulcers of the cornea, sometimes dendritic ulcers, or as vesicles on the eyelids. Later chorioretinitis may develop. The cornea may develop opacifications after recurrence, indicating a progressive involvement. Blindness may be the consequence. Even herpetic necrosis of the retina has been observed as a very rare consequence of the infection.

THE VIRUS

These viruses are members of the *Alphaherpesvirinae* subfamily of human herpesviruses together with varizella-zoster virus, also called human herpesvirus 3 (Figure 18.2). HSV1 and HSV2 have a wide host-cell range in contrast to other members of the family. All herpesviruses establish lifelong persistent infections which may reactivate after short or long time intervals. In immunosuppressed individuals frequent or severe reactivations of HSV may cause a considerable problem. Latent infections are localized to neurons of the sensory ganglia, and may cause reactivation often quite different from the primary disease. HSV is a large virus with a core containing double-stranded DNA within a coat, an icosahedron with 162 capsomeres. The envelope which surrounds the 'naked' particle is partly nuclear membrane derived, partly virally coded, with glycoprotein spikes. The diameter of a complete particle is 120–200 nm. The 'naked' virion measures about 100 nm.

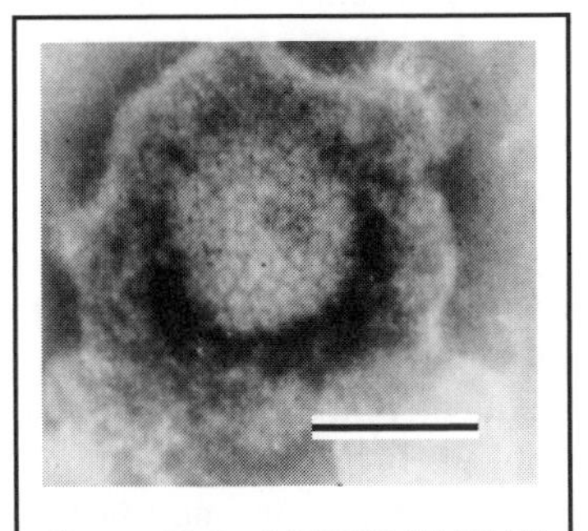

Figure 18.2 HERPES SIMPLEX VIRUS. Bar, 100 nm (Electron micrograph courtesy of E. Kjeldsberg)

The virus enters the cells through cellular membrane fusion after being attached to specific receptors through an envelope glycoprotein. The capsid is transported to nuclear pores where DNA circularizes after uncoating, and enters the nucleus. The genome is a 120–230 kbp double-stranded, linear DNA with repeated sequences located at each flank and certain other regions. Genomic rearrangements giving genomic 'isomers' may occur, but the biological significance of this is unknown. HSV1 and HSV2 show 50% sequence homology.

'Fingerprinting' of separated restriction enzyme digests shows differences between HSV1 and HSV2, and between strains within each type. This may allow epidemiological tracing. Expression of the genome is strictly regulated, and occurs in a certain order. First the immediate early genes start yielding so-called 'alpha' proteins which induce production of 'beta' proteins by permitting the expression of an early set of genes. The viral DNA starts to replicate, and the late 'gamma' proteins, mostly structural proteins, are produced. There are 35 structural proteins, and the genome may code for at least 65 additional proteins, e.g. viral enzymes not incorporated into the virion.

Viral DNA is synthesized by the help of a large number of viral enzymes which may be targets for antiviral drugs. The newly synthesized viral genomes are packed into empty nucleocapsids in the cell nucleus. The synthesis of viral RNA is possible through the replicative cycle of viral DNA using cellular RNA polymerase II with viral factors participating.

The filled nucleocapsids bud through an altered inner nuclear membrane. Viral particles can then be released through the cellular membrane or through vacuoles on the surface as an enveloped virus. The herpes virus has an 18 hour replication cycle. Cellular macromolecular synthesis is shut off early during the replicative cycle, leading to cell death.

EPIDEMIOLOGY

HSV infections exhibit no clear seasonal variations, and the virus is maintained in man by latent and recurrent infections. The only reservoir is in man. Viral excretion is highest in persons with active lesions, but excretion may occur in up to 15% of asymptomatic individuals. HSV2 antibodies are less frequent in nuns (3%), while they are highest in prostitutes (70%). HSV2 antibodies are more prevalent in individuals belonging to lower socioeconomic classes and among promiscuous groups.

The frequency of recurrence shows individual variations, probably affected by other infections, sunburn, menstruation or trauma in general.

HSV2 genital infection reactivates more frequently than HSV1, and frequently recurring genital herpes is a major problem. It has been estimated that the number of recurrences is more than 8–9 times annually in one-third of the cases, less than 2–3 each year for another third and intermediate for the rest.

Neonatal herpes occurs in about 1 in 5000 deliveries (USA), in some areas even higher. The majority of children (75%) appear to be infected during passage through the genital tract. The overall risk of neonatal infection, however, is 10 times greater in primary genital infection of the mother than if the infection is reactivated. The severity of the disease is apparently not dependent on the type of HSV, if the child is premature or full-term, or if the disease has been contracted postpartum or at delivery. However, the risk of being infected is higher in prematures, in babies delivered with instrumentation (e.g. scalp electrodes), if cervix is involved at delivery, and in primary infection of the mother. Many studies do not support any link between HSV and carcinoma of the cervix.

THERAPY AND PROPHYLAXIS

Primary HSV infections, or relapses, can be treated with aciclovir (acycloguanosine) which reduces the severity and duration of symptoms.

However, elimination of latent HSV in sensory ganglia cannot be obtained using any treatment, only the number of recurrences can be reduced by systemic aciclovir treatment. The most effective dosage seems to be 200 mg q.i.d.

HSV encephalitis is also treated with aciclovir. Treatment should start before development of coma in order to be fully effective. In neonatal herpes aciclovir will reduce the mortality rate. In immunosuppressed patients systemic aciclovir will suppress reactivation or shorten the duration of viral excretion in recurrent disease, and reduce local pain and time of healing. Herpetic keratitis has been treated with topically applied trifluoridine (trifluorothymidine), but idoxuridine (5-iodo-2′-deoxyuridine), vidarabine and aciclovir are also effective. Herpetic stromal disease and iritis may effectively be treated with a combination of corticosteroid and antiviral without additional risk. Systemic aciclovir may prevent frequent relapses of ocular herpes. If the HSV is resistant to aciclovir, vidarabine has been recommended for treatment of encephalitis or neonatal herpes, and brivudin for treatment of HSV1 in immunosuppressed patients or in eye infections. Foscarnet may be effective in aciclovir-resistant, thymidine kinase-deficient HSV infections, and is recommended by some as the drug of choice in treating infections with resistant HSV. Other antiviral agents considered for use in herpesvirus infections include brovavir, penciclovir, desciclovir, bishydroxymethylcyclobutylguanine (BHCG), 1-(3-hydroxy-2-phosphonylmethoxypropyl)cytosine (HPMPC) and, in particular, famciclovir and valaciclovir.

Especially the primary infection of expecting mothers at term requires caesarean section in order to protect the child. This should be done in advance of rupture of the membranes. Otherwise, prophylactic aciclovir may be indicated for the baby. Viral cultivation seems of little value as a basis for the decision to perform caesarean section. However, visible herpetic lesions in the genital area, whether primary or recurrent, are a useful guide. Current opinion is that transplacental transmission occurs infrequently (1 to 2 in 200,000 deliveries in the USA), and no prophylactic measures are available during pregnancy.

Even though **vaccine** against chickenpox has been used for years, HSV vaccines are still not available.

LABORATORY DIAGNOSIS

Isolation in tissue culture is still useful. Electron microscopy or immunofluorescence is also useful in early cultures, and immunofluorescence when studying infected cells directly from the patient. Serology is of no use in recurrences except in cases of encephalitis.

The encephalitic lesion may be visible early on in images made by computerized tomography (CT) or magnetic resonance imaging (MRI), and develops faster than a (malignant) neoplastic tumour. The aetiological diagnosis can be established by biopsy of affected brain tissue, but this procedure has been replaced by indirect methods such as tests for local antibody production and HSV PCR in CSF. Production of antibody to HSV within the brain starts as early as 2–3 days after clinical symptoms in a small fraction of patients. However, this fraction increases to nearly 100% over the next 2 weeks. A possible serological cross-reaction between HSV and VZV is of minor clinical importance as the treatment would be the same.

The HSV PCR, using prepared CSF as template, is sensitive and will be positive early in the disease. After 5–6 days the sensitivity falls, probably as a result of the removal of virus by local anti-HSV antibodies. Thus, HSV PCR and CSF antibody tests complement each other. The PCR method can, by using specific probes, also identify the viral type.

Diseases mimicking oral herpes infections are other vesicular or ulcerating lesions: herpangina caused by coxsackie virus, mononucleosis by Epstein–Barr virus, Stevens–Johnson syndrome, aphthous stomatitis, bacterial infections or lesions caused by drug intolerance, irradiation or immunosuppressive therapy. The diagnosis of genital HSV infection should include chancroid, syphilis, genital lesions, Behçet syndrome, erythema multiforme, local candidiasis and simple erosions. The symptoms of neonatal herpes can be caused by other infections such as rubella, cytomegalovirus and toxoplasma, and sometimes by erythroblastosis.

CHICKENPOX—THE FIRST ENCOUNTER

19. VARICELLA-ZOSTER VIRUS (VZV)—VARICELLA

Chickenpox. Lat. *varicella* = vesicle; Ger. *Wasserpocken, Windpocken*; Fr. *varicelle, petite vérole volante.*

A. Winsnes and R. Winsnes

Primary infection with VZV causes varicella (chickenpox), a common, usually mild exanthematous childhood disease. The infection is more serious in adults and may be prolonged and life-threatening in newborns and individuals with immunodeficiency. Zoster, 'herpes zoster', is a clinical manifestation of reactivation of latent VZV, see Chapter 20.

TRANSMISSION/INCUBATION PERIOD/CLINICAL FEATURES

Varicella is most infectious 2 days before and 3–4 days after the eruption. VZV is spread by direct contact or by droplets, but may also be airborne in institutions. The incubation period is usually 14–16 days.

SYMPTOMS AND SIGNS

Systemic:	Variable Fever and Mild General Malaise for 2–4 days
Local:	Vesicular, Pruritic Rash on Mucous Membranes and Skin

COMPLICATIONS

Secondary staphylococcal or streptococcal skin infection with impetigo or erysipelas is common. More rarely invasive infection may lead to septicaemia and distant purulent foci such as arthritis or osteomyelitis. Primary VZV infection may spread to all internal organs and commonly causes pneumonitis in adults. Serious and potentially lethal VZV infection is frequently seen in immunocompromised patients.

THERAPY AND PROPHYLAXIS

When given within 24 hours after the eruption antiviral therapy with aciclovir is effective in shortening the duration of varicella. Famciclovir and valaciclovir are also effective and have better bioavailability. Foscarnet is used in the seldom aciclovir-resistant cases. Antiviral therapy is recommended in infants, adults and immunocompromised patients. Specific VZ-immunoglobulin (VZIG) given up to 3 days after VZV exposure is usually protective, but is reserved for use mainly in high-risk patients. High-risk patients can also be protected by vaccination.

LABORATORY DIAGNOSIS

The diagnosis is clinical given by the typical vesicular rash, and laboratory confirmation is seldom necessary. If the rash is atypical direct fluorescent antibody staining of cell scrapings or identification of virus antigen by PCR technique is most useful. PCR analysis of cerebrospinal fluid is valuable if neurological complications occur. A rising antibody titre in paired serum samples is diagnostic. Serologic testing may be unreliable in immunocompromised patients.

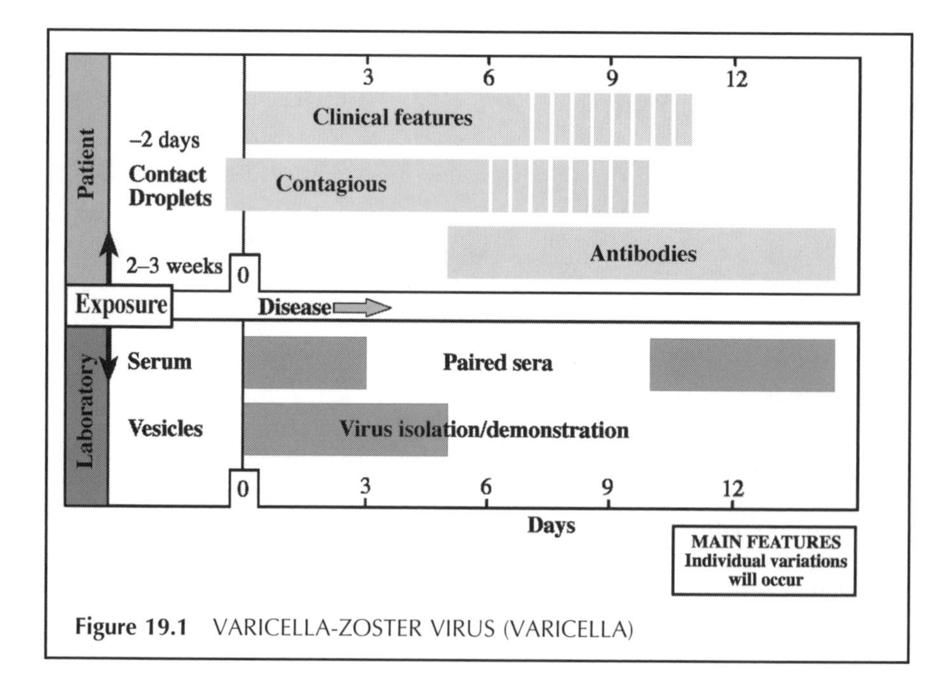

Figure 19.1 VARICELLA-ZOSTER VIRUS (VARICELLA)

SYMPTOMS AND SIGNS

The incubation period is usually 14–16 days, but may vary from 10 to 21 days, and up to 28 days after prophylactic treatment with VZIG. Especially in teenagers and adults, prodromal symptoms with malaise and low-grade fever may occur 1–2 days before eruption of the vesicular rash. The typical crops of varicella lesions initially observed on the face, scalp or trunk develop during hours from pruritic macules to oval 2–3 mm vesicles with clear fluid becoming cloudy and crusty after 1–2 days. During the first 2–4 days skin lesions at different developmental stages coexist. Usually 100–300 lesions are found, and the active disease lasts for 1 week. The diagnosis may be missed when only a few lesions occur, and rarely the infection may be subclinical. Mucosal vesicles in the mouth, pharynx, conjunctiva and the external genitalia rupture easily and may therefore be overlooked. Itching is common and usually mild constitutional symptoms and fever occur during the first days of the rash. Higher fever and more symptoms generally accompany an extensive eruption.

Differential diagnosis. Mild cases of varicella may go unnoticed or be mistaken for impetigo. In hand, foot and mouth disease due to coxsackie virus, vesicles up to 5 mm occur on hands, feet and mouth mucosa, but not on the trunk (Chapter 8). Seldom will herpes simplex viral infection in children with atopic eczema become varicella-like (Chapter 18). Very rarely a varicelliform rash is caused by *Rickettsia akari* transmitted by a mouse mite. These vesicles are smaller, more deeply seated on a firm papule, and lack the typical crusting seen in varicella.

CLINICAL COURSE

In healthy children varicella usually has a benign course. Secondary family contacts often have a more severe disease due to a higher viral load. Extensive eruptions and a more serious course may be seen in adults. If the mother gets varicella during the perinatal period, especially 5 days before or 2 days after delivery, the infant may develop serious varicella, often fatal if untreated, due to lack of maternal antibody protection. Prematures born before 30 weeks' gestation also lack maternal antibodies. Other high-risk groups for potential fatal VZV infection are patients with compromised cellular immunity such as leukaemia, lymphoproliferative diseases, HIV infection and individuals treated, for whatever reason, with corticosteroids and other immunosuppressive or cytotoxic drugs. VZV infection usually results in lifelong immunity against reinfection, but not against reactivation (zoster).

COMPLICATIONS

Common complications are superinfection with *Staphylococcus aureus* or group A *Streptococcus pyogenes* causing impetigo and erysipelas or less commonly extensive cellulitis, necrotic or bullous skin infection. Bullous varicella is caused by epidermolytic toxin-producing staphylococci. Invasive infection with septicaemia, arthritis, osteomyelitis or bacterial pneumonia may occur.

Visceral spread of VZV can affect the lungs, brain, liver, pancreas, kidneys or heart. Fatal cases are most often due to interstitial varicella pneumonia (pneumonitis) which is 25 times more common in adults than in children. Smokers, pregnant women and patients with chronic lung diseases are at increased risk for developing serious pneumonitis that may be fatal. Usually rapid clinical recovery takes place, though radiographic changes may persist for weeks, sometimes leaving calcifications.

Neurologic complications include meningoencephalitis due to direct VZV infection during the first week with high fever and deterioration of consciousness. VZV may also play some direct part in cerebellar ataxia, thought to be mainly immunological. Ataxia usually starts 1 week after appearance of the rash and is a benign condition lasting up to 1–2 weeks in children. Rare cases of limb paresis due to transverse myelopathy or brain arteritis have been reported. Reye syndrome (fatty liver and encephalopathy) has become very rare since the use of aspirin has declined. VZV hepatitis, mostly subclinical, still occurs as a separate entity.

Rarely thrombocytopenia, haemorrhagic varicella, fulminant purpura and leucopenia may occur, especially in patients with immunodeficiency. A congenital varicella syndrome with limb atrophy and scarring of the skin occurs after VZV infection in 1–2% of those contracting varicella during the first 20 weeks of pregnancy.

THE VIRUS

Varicella-zoster virus (Figure 19.2) is one of eight herpesviruses. It is a double-stranded DNA virus 150–200 nm in diameter with a lipid envelope where glycoprotein spikes surround an inner icosahedral nucleocapsid. After penetration of the infected cell, the virion is uncoated, and the capsid penetrates the cell nucleus where replication occurs. Viral DNA is integrated in the host cells thereby avoiding immune surveillance and eradication by antiviral drugs. VZV is quickly inactivated outside host cells. Haematogenous spread by mononuclear cells, secondary viraemia, occurs 4–5 days before and 1–2 days after onset of symptoms. Man is the only natural host. Only one antigenic VZV type has been identified. Attenuated viral strains have been developed through serial passages in cell cultures, and the Oka strain is used in the live-virus vaccine now available. Mutant VZV strains resistant to aciclovir have been isolated, especially from AIDS patients repeatedly treated with antiviral drugs.

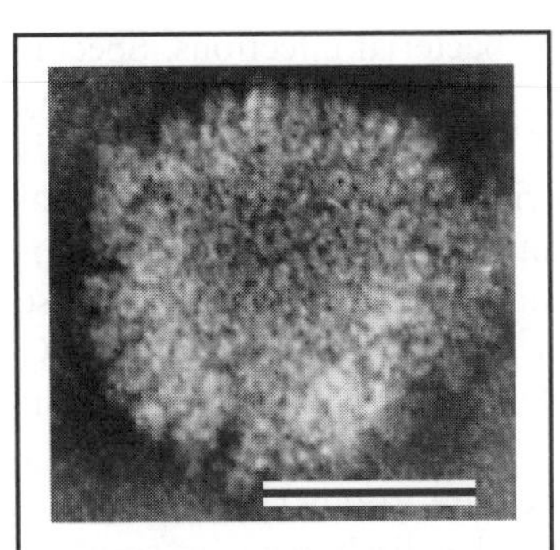

Figure 19.2 INNER NUCLEOCAPSID OF VARICELLA-ZOSTER VIRUS. Bar, 50 nm (Electron micrograph courtesy of G. Haukenes)

EPIDEMIOLOGY

Varicella is very contagious. Thus 90% of susceptible household contacts contract the disease. Varicella is therefore predominantly a childhood disease in temperate areas where 90% of cases occur below 10 years. In the USA 96% of adults are immune, while adults often remain susceptible to varicella in tropical countries. Epidemics are seen in temperate climates most frequently during late winter and early spring. Infection is usually spread by droplet or direct contact, but may be airborne in institutions. The infectivity is maximal 1–2 days before and 3–4 days after the eruption, but may be extended if new crops of vesicles occur. Nosocomial infection is a serious problem, especially in units treating malignancies and immunodeficient patients and performing transplantations.

THERAPY AND PROPHYLAXIS

The antiviral drug aciclovir has improved the prognosis of serious VZV infections. VZV is, however, less sensitive than HSV, and therefore a 4-fold higher dose is needed against VZV. Antiviral therapy is not recommended for use in children without chronic disease, except secondary household and teenage cases. Risk groups such as adults and immunocompromised individuals should be offered treatment, preferably intravenously. Antiviral treatment should also be given when complications such as varicella pneumonia and encephalitis occur.

Aciclovir and penciclovir with their respective oral prodrugs valciclovir and famciclovir reduce clinical symptoms and shorten the course of VZV infection when started within 48 hours after skin eruption. Early treatment gives the best results. Hopefully the occurrence of complications is reduced, though this has not been proved due to their rarity. These antiviral drugs are all dependent on the virus-encoded thymidine kinase for intracellular activation. Cross-resistance to these drugs has been reported for viral strains isolated from AIDS patients having had repeated treatment courses with aciclovir. When VZV resistance is suspected, treatment with foscarnet should be given.

Specific immunoglobulin has no proven therapeutic effect. In uncomplicated varicella symptomatic treatment of pruritus is recommended to prevent impetigo. Antibiotics are given against secondary bacterial infections. Specific zoster immunoglobulin (VZIG) given up to 3 days after exposure may prevent or modify clinical disease. Because of the scarcity of VZIG, this preparation must be reserved for use in high-risk patients. Pooled normal immunoglobulin preparations contain small amounts of specific immunoglobulin, insufficient to prevent disease in ordinary doses. Probably high-dose immunoglobulin given intravenously exerts a prophylactic effect. VZIG treatment does not prevent the development of immunity unless the patient has an immunopathy.

In hospitals strict isolation (in negative-pressure rooms) of infectious patients is necessary, or preferably, they should be discharged as soon as possible and treated as outpatients.

Varicella vaccine with live attenuated VZV has proven effective in protecting healthy individuals as well as high-risk patients against varicella. Seroconversion rates after one vaccine dose are at least 95% in children younger than 12 years, whereas older persons require two doses for equivalent protection. Non-immune individuals scheduled for transplantation should be vaccinated at least 3 months before operation. In most countries vaccination is recommended for use in non-immune teenagers and adults, whereas widespread vaccination of healthy young children is not recommended, though the vaccine may be approved for administration to this group. Vaccination during ongoing cytostatic treatment is less effective, and usually not recommended 6 months after postponing such treatment.

LABORATORY DIAGNOSIS

Usually the clinical diagnosis is accurate with no need for laboratory tests. For diagnostic help in the acute stage the sensitive PCR technique can detect VZV DNA in vesicles, blood and spinal fluid. VZV is abundant in vesicle fluid, but the electron microscopic picture cannot be distinguished from HSV and CMV. Immunofluorescent staining of vesicle fluid with monoclonal antibodies can identify VZV. The cytopathic effect of VZV in cell culture is characteristic, but takes some time, and VZV is not always readily cultured. A rise in antibody titre or demonstration of specific IgM usually confirms the diagnosis. However, during VZV infection a simultaneous antibody rise against HSV may occur, and vice versa. Specific CF antibodies are found 6–7 days after the onset of the rash. CF-antibody titres may, however, be below detectable level 3 years after the infection. Then latex agglutination assay, indirect immunofluorescence or ELISA techniques are necessary to detect VZV antibodies. Because of the potential severe course of VZV infection, and the possibility of giving prophylaxis by exposure, sensitive techniques are needed for identification of susceptible persons.

HELL'S FIRE—CHICKENPOX REVISITED

20. VARICELLA-ZOSTER VIRUS (VZV)—ZOSTER

Shingles. Gr. *zoster* = belt; Ger. *Gürtelflechte*, *Gürtelrose*; Fr. *zona*.

A. Winsnes and R. Winsnes

Zoster, 'herpes zoster', is usually a disease of adults, especially elderly people, caused by reactivation of latent varicella-zoster virus (VZV) in dorsal root ganglion cells. In AIDS and other immunocompromised patients zoster is both a frequent and dreaded disease.

REACTIVATION/TRANSMISSION

VZV persists in a latent form in sensory nerve cells (dorsal roots of the spinal medulla or cranial nerve ganglia) for decades after varicella infection. Though re-exposure to VZV may be a factor in reactivation of virus in some circumstances, it is generally poorly understood why VZV starts replicating and spreading down sensory nerve fibres. VZV appears in vesicles on the skin area corresponding to the dermatome innervated by the nerve in question. Although zoster is less contagious than varicella, it may cause varicella in susceptible contacts.

CLINICAL FEATURES

Neuralgic pain and tenderness in the affected area frequently start several days before eruption of the rash. The zoster vesicles are usually somewhat larger than those of varicella. The development to crusting is slower (7–10 days), and the occurrence of new crops of vesicles is seen less often than in varicella. Pigmentary changes and scarring may be seen following the loss of crusts after 3–4 weeks. The rash is usually unilateral and localized to the area (dermatome) innervated by one or two sensory nerves. Localization is most frequent on the thorax, neck or face. With involvement of cranial nerves vesicles may occur on the eyes, in the external ear canal and in the mouth. Regional lymph nodes are regularly enlarged and tender. General symptoms with malaise and fever are usually not very prominent. The uncomplicated clinical course is 1–3 weeks.

COMPLICATIONS

Complications are especially seen when zoster is located in cranial nerve areas or when the host resistance is compromised. Involvement of the ophthalmic branch of the trigeminal nerve (zoster ophthalmicus) may result in dendritic keratitis that may cause scarring of cornea and reduced vision. VZV may also cause retinitis with poor visual prognosis. Immediate ophthalmological examination is recommended. Involvement of the seventh cranial nerve may cause facial nerve palsy (Ramsay Hunt syndrome), where prognosis for recovery is not so good as in the common Bells palsy. Pareses are due to spread of the virus to the motor neurons in the medulla or cranial nerve ganglia. By EMG it has been shown that motor involvement occurs in 35% of the thoracic zoster cases.

Other neurologic complications such as encephalitis, myelitis and polyneuropathy may be caused by immunological inflammatory processes, but also by direct spread of VZV, even without the presence of the typical zoster rash. PCR methods have in several cases shown VZV DNA to be present in mononuclear blood cells, blood vessels and spinal fluid. Thus VZV may cause neuronal damage because of direct destruction of neurons and compromised blood flow because of arteritis in small or large vessels.

In zoster patients who are 50 years and older postherpetic neuralgia (pain persisting more than 6 weeks after appearance of the rash) is a common complication, occurring in 40% of zoster patients above 60 years. VZV DNA has been detected in mononuclear blood cells of some patients with postherpetic neuralgia, and speculation of a possible higher viral load in patients with neuralgia would argue for aggressive antiviral treatment. Once postherpetic neuralgia disappears, it does not recur.

In immunocompromised patients, particularly transplant recipients, cancer and AIDS patients, zoster may become generalized and life-threatening. Haematogenous dissemination to internal organs may occur as in varicella (Chapter 19). As in varicella secondary bacterial infections of the rash may occur, sometimes becoming invasive.

EPIDEMIOLOGY

With increasing age cellular immunity becomes weaker, explaining why zoster is 10 times more frequent in persons over 70 years of age than in teenagers. The disease is also more prevalent and serious among immunocompromised patients. It has been calculated that by the age of 80 about 50% will have had one attack of zoster, and 1% in this age group will have had two attacks. Increased risk of contracting zoster is seen in children who have had varicella infection during fetal life or early infancy, probably due to lower specific immunity. Adults with frequent re-exposure to varicella through contact with children have a lower incidence of zoster.

THERAPY AND PROPHYLAXIS

When given within the first 3–5 days after eruption of the rash, aciclovir has proved effective for treatment of zoster both in otherwise healthy and immunocompromised patients. In the latter group intravenous antiviral treatment shuld be given as soon as possible. In acute zoster the standard treatment is 7–10 days of oral aciclovir treatment. For ophthalmic zoster, topical treatment with aciclovir is given in addition to systemic treatment. A variety of treatment regimens (even combined treatment with aciclovir and prednisolone) against postherpetic neuralgia have had limited success. Possibly more aggressive antiviral treatment at the start of zoster will diminish the occurrence of complications, but this is not settled so far.

LABORATORY DIAGNOSIS

As zoster is caused by reactivation of VZV, the IgG response in serum is quicker and more pronounced than that seen in varicella. Specific IgM is found in small amounts. Viral DNA may be identified by PCR methods used with vesicular fluid or scrapings, as well as with blood and spinal fluid. VZV infection of the nervous system may be protracted, especially in immunocompromised patients. In dermatomal pain without rash (preherpetic zoster and 'zoster sine herpete'), and in cases of acute pareses or meningoencephalitis or myelitis, PCR analyses may be important.

WE HAVE ENOUGH PROBLEMS HERE WITHOUT YOU

21. CYTOMEGALOVIRUS (CMV)

The name of the virus refers to the size of the infected cells, which contain large intranuclear inclusions.

A. B. Dalen

The cytomegaloviruses (CMV) belong to the herpesvirus family (subfamily *Betaherpesvirinae*). They are widely distributed in man and other mammals, but possess a high degree of species specificity. Human infections may be asymptomatic, or cause severe, generalized disease. Following primary infection, CMV establishes a latent infection of lymphocytes and possibly endothelial cells from which they may be reactivated.

TRANSMISSION/INCUBATION PERIOD/CLINICAL FEATURES

Fetal infection can follow reactivation or primary maternal infection. Perinatal infection occurs through infected cervical secretions and milk. Postnatal infection is acquired by the respiratory route, infected semen and through blood transfusions and organ transplants. The incubation period is about 3–6 weeks.

SYMPTOMS AND SIGNS

Postnatal:	Infectious Mononucleosis-like
Congenital and immuno-compromised patients:	Extensive Organ Damage in Severe Cases

Most infections are asymptomatic. After an insidious start symptoms may last 1–5 weeks. Immunocompromised patients: Extensive organ damage in severe cases.

COMPLICATIONS

Interstitial pneumonia, hepatitis and occasionally Guillain–Barré syndrome. Retinitis, gastrointestinal infection.

THERAPY AND PROPHYLAXIS

Foscarnet and ganciclovir. CMV vaccines are still at the developmental stage.

LABORATORY DIAGNOSIS

Virus can be cultured from urine, saliva, blood, milk, cervical discharges, semen, biopsies and autopsied organs. Laboratories usually receive urine, blood or bronchoalveolar lavage samples. Immunocytochemical assays for CMV may be performed on the same materials. The PCR technique is widely used in detecting CMV infections. A quantitative PCR or additional tests for CMV are usually required to establish an aetiological diagnosis. Tests are available for CMV IgM and IgG antibodies. A latex agglutination test is available for rapid IgG antibody screening of blood donors.

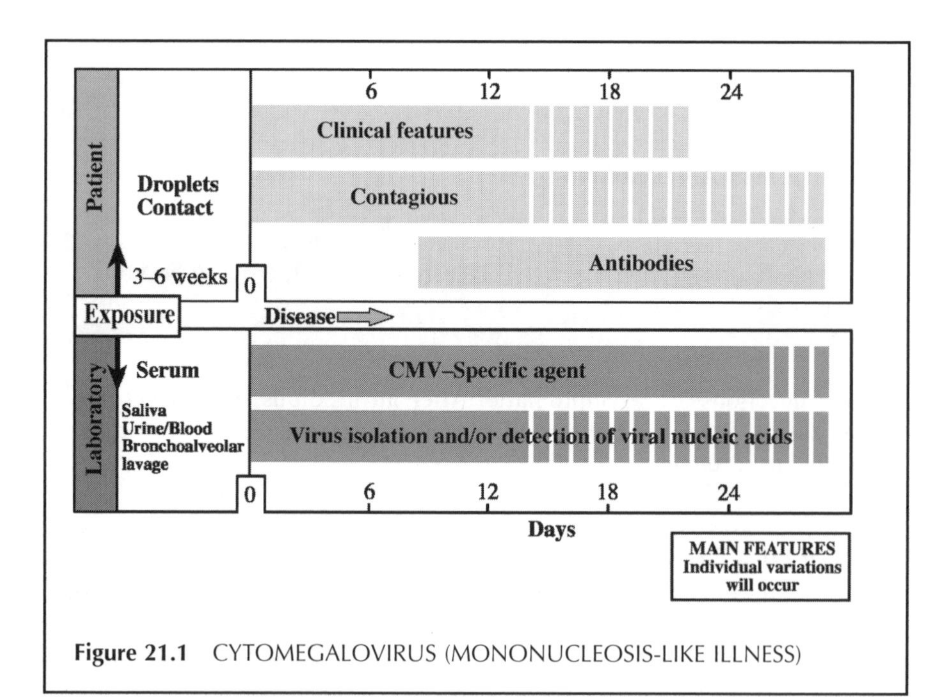

Figure 21.1 CYTOMEGALOVIRUS (MONONUCLEOSIS-LIKE ILLNESS)

CLINICAL FEATURES

SYMPTOMS AND SIGNS

CMV is an opportunistic agent which may seriously damage a developing fetus, while disease is otherwise rare unless the host has lowered resistance to infection. Reactivation of latent CMV is also often related to immune deficiencies.

Congenital infection. CMV can be transmitted *in utero* in both primary and reactivated maternal infections. Gestational age at the time of maternal infection does not seem to affect transmission *in utero* or expression of disease in the fetus. Infants have a generalized infection at birth, but only 5–10% of them have clinical symptoms at birth. Generalized cytomegalic inclusion disease of the newborn results mostly from primary maternal infection. The most common signs are in decreasing order of frequency: petechiae, hepatosplenomegaly, jaundice, microcephaly and chorioretinitis. Multiple organ involvement is frequent. The fetal infection may show minimal manifestations at birth and still result in significant damage in later life, especially to the central nervous system. Infants with subclinical infections at birth may develop sensorineural deafness within the first years of life. Minimal brain dysfunction syndromes have been reported in children with a congenital, subclinical CMV infection not apparent at birth.

Perinatal infection. Newborn infants become infected from exposure to CMV in cervical secretions of the mother at delivery or in breast milk 2–4 months post partum. Perinatal CMV infections are subclinical with the exception of rare pneumonias.

Postnatal infection. The incubation period is 3–6 weeks following transfusion, and may be longer after naturally acquired infection. Salivary spread is common and sexual transmission may occur. The infections are usually subclinical, but infectious mononucleosis may occur. The disease is characterized by malaise, myalgia, protracted fever and liver function abnormalities. Atypical, peripheral lymphocytes may resemble those of EBV mononucleosis. Lymphadenopathy is usually not prominent, and heterophile antibodies are not present. Reactivation of CMV is consistently seen in seropositive patients following renal transplantation. Clinical symptoms in primary infections through transfusions or latently infected donated organs are seen in about 85% of transplant recipients with primary infection and in 20–40% of those with a recurrent infection. The most common sites of involvement are: adrenals, lungs, gastrointestinal tract, CNS and eyes (retinitis). In acquired immunodeficiency due to infections (AIDS) or immunosuppressive regimens both recurrent and primary infections with a high morbidity occur with high frequency.

Differential diagnosis. Congenital CMV infections must be distinguished from congenital rubella and toxoplasmosis by laboratory means. CMV mononucleosis closely resembles EBV mononucleosis clinically, but can be differentiated by the lack of heterophile antibodies and the application of CMV antibody tests. CMV infections in immunocompromised individuals are often complicated by other infections which make the clinical diagnosis difficult.

COMPLICATIONS

Signs of hepatitis are often seen, but it is usually mild and never becomes chronic. The interstitial pneumonia that may develop in immunocompromised patients is severe and life-threatening. CMV infections occasionally seem to be associated with the Guillain–Barré syndrome.

THE VIRUS

CMV (Figure 21.2), belonging to the *Betaherpesvirinae* subfamily, is the largest of the members of the human herpesvirus family (200 nm in diameter). The morphology is similar to that of other members of the group with a 64 nm core containing double-stranded viral DNA, enclosed by a 110 nm icosahedral capsid and an outer envelope. Human CMV is strictly species-specific and infects cell cultures of fibroblasts and to a lesser extent certain epithelial cells and B- and T-lymphocytes. Latent infections *in vivo* are found in leucocytes and possibly endothelial cells. A great number of genetic variants of CMV have been demonstrated by the use of restriction endonuclease assays. There is at present no generally accepted immunological system for classification of CMV. The virus contains 33 structural proteins and codes for an unknown number of non-structural proteins. The glycoproteins of the envelope are important antigens. The cytopathic effect in tissue culture characteristically consists of islands of enlarged cells with nuclei filled up with large inclusion bodies.

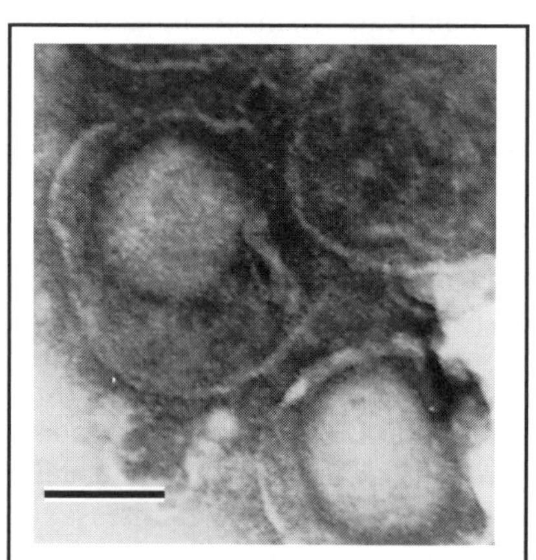

Figure 21.2 CYTOMEGALOVIRUS. Bar, 100 nm (Electron micrograph courtesy of A. B. Dalen)

EPIDEMIOLOGY

Human CMV is ubiquitous and humans are the only reservoir. The virus is readily inactivated, and close contact is required for horizontal spread. The sources of virus are oropharyngeal, vaginal and cervical secretions, semen, urine, breast milk and blood. Infection rates vary greatly, and socioeconomic factors affect both intrauterine and extrauterine transmission. By puberty, 40–80% of children have been infected, and the prevalence increases to 70–90% in adults. Cervical shedding of CMV is common in pregnancy, 1–2% in the first trimester, 5–10% in the second and 10–15% in the third. The risk of congenital infections varies between 0.4 and 2.6% (average 1%) of live births. Of seropositive mothers, 30% or more excrete CMV intermittently into breast milk, most commonly 2–4 months post partum. Intrauterine, perinatal and early postnatal acquisition of CMV is followed by a prolonged excretion of virus (5 years or more in the urine, 2–4 years in the nasopharynx).

THERAPY AND PROPHYLAXIS

Foscarnet, a pyrophosphate analogue and ganciclovir, a nucleoside analogue with a modified pentose, have anti-CMV activity and are in clinical use both in prophylaxis (transplantations) and in suppressive treatment in established infections. Foscarnet has a survival benefit compared to ganciclovir for CMV retinitis in AIDS patients, but may be less well tolerated. The prophylactic use of human leucocyte interferon and human hyperimmune gammaglobulin have had limited success. CMV vaccines are still at the developmental stage. Practical measures such as the use of CMV seronegative blood for immunocompromised individuals requiring frequent transfusions (especially prematures) reduce the incidence of severe CMV infections.

LABORATORY DIAGNOSIS

Virus isolation, which is the preferred method for diagnosing productive CMV infections, is performed on urine and/or throat or genital secretions, milk and blood. The sample should be mixed with a suitable transport medium, brought to the laboratory as soon as possible and never be frozen. CMV is grown in human lung fibroblasts, and the use of monoclonal antibodies allows an early detection of virus-infected cells (24–48 hours). Monoclonal antibodies may be used for the detection of CMV in biopsy material, blood leucocytes or cells obtained by bronchoalveolar lavage. Tissues should be brought to the laboratory in transport medium or fixed in ethanol. Blood leucocytes may be obtained from buffy coats. Positive findings are highly indicative of productive infections. The PCR method is extremely sensitive and a positive reaction may only reflect a latent infection. A (semi)quantitative PCR is therefore highly desirable. Sensitive ELISA allows the detection of specific IgG and IgM antibodies. IgM antibodies peak early in the infection and are usually undetectable 12–16 weeks after the onset of subclinical infections. IgM persists for longer periods in symptomatic infections and especially in congenital infections. Low levels of IgM antibodies may be detected in recurrent CMV infections. Specific CMV serum IgG antibodies last for decades. A variable rise in the IgG titre is seen in recurrent CMV infections. It is difficult to distinguish primary from recurrent CMV infection in immunocompromised patients. A pretreatment serological status is of great value in transplant patients and cancer patients receiving chemotherapy.

BEWARE YOUR SUITORS

22. EPSTEIN–BARR VIRUS (EBV)

M. A. Epstein and Y. M. Barr: Scientists who discovered the virus in 1964.
Infectious mononucleosis (IM); glandular fever; kissing disease.

E. Tjøtta

Epstein–Barr virus belongs to the herpesvirus group. After the primary infection the virus establishes a lifelong latency. Nearly all adults have been infected with EBV.

TRANSMISSION/INCUBATION PERIOD/CLINICAL FEATURES

Infectious virus can be isolated from saliva for months after primary infection and can be transferred by kissing. The incubation period has been estimated to be between 30 and 50 days, sometimes longer.

SYMPTOMS AND SIGNS

Local:	Tonsillitis
Systemic:	Fever, Fatigue, Lymphocytosis with Atypical Lymphocytes, Lymphadenopathy, Splenomegaly
Other:	*See* Complications

Symptoms last for about 10 days (1–4 weeks). Most infections in children are asymptomatic. Pronounced fatigue and protracted convalescence sometimes occur in adolescents and adults.

COMPLICATIONS

Allergic rash following use of antibiotics, especially ampicillin, is common. Rare complications are rupture of the spleen, airway obstruction, fatal myocarditis, meningoencephalitis, icterus, nephritis, pneumonia, thrombocytopenic purpura and Guillain–Barré syndrome. In some patients a chronic fatigue syndrome has been linked to EBV infection. EBV may cause special problems in immunodeficient persons. EBV may be associated with certain malignant tumours and also with erythrophagocytosis.

THERAPY AND PROPHYLAXIS

Usually no specific treatment or prophylaxis.

LABORATORY DIAGNOSIS

Leucocytosis with atypical mononuclear cells occurs in the second week of illness. The presence of heterophile antibodies with a particular absorption pattern is diagnostic of glandular fever (Paul–Bunnell test). Specific EBV antibodies appear early and some persist for years.

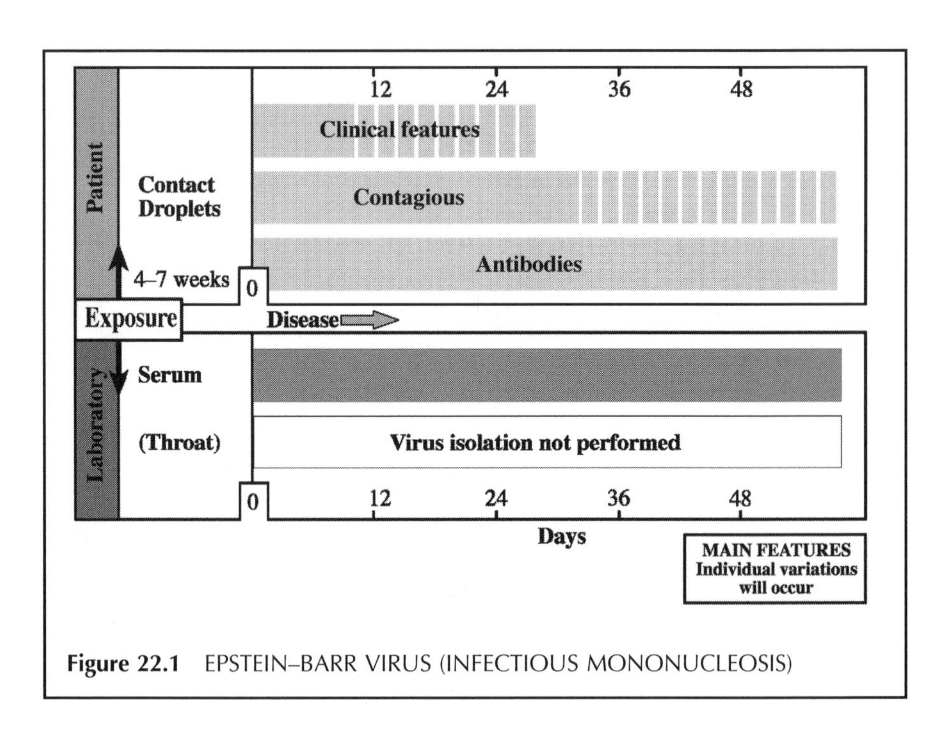

Figure 22.1 EPSTEIN–BARR VIRUS (INFECTIOUS MONONUCLEOSIS)

CLINICAL FEATURES

The virus is mainly shed from oropharynx, but may be excreted in the uterine cervix. Replication has been found mainly in B-lymphocytes, but also in epithelial cells in the pharynx, parotis and cervix, as well as in epidermal skin of immunocompromised patients. The typical primary illness is infectious mononucleosis (IM). Atypical illnesses are meningoencephalitis, myocarditis, and hepatitis. EBV has also been linked to hairy leukoplakia of the tongue in patients with AIDS and in malignancies such as Burkitt's lymphoma, nasopharyngeal carcinoma, Hodgkin's lymphoma, lethal midline granuloma, thymoma and malignant lymphoepithelial lesions of the salivary gland.

Primary infection	Reactivated EBV infection
Asymptomatic	Chronic active EBV infection
Infectious mononucleosis (IM)	Lymphoproliferative disorders
Primary atypical EBV infection	Burkitt's lymphoma
X-linked lymphoproliferative syndrome	Nasopharyngeal carcinoma
Meningoencephalitis	Other tumours
Myocarditis	Interstitial pneumonia
Hepatitis	Uveitis
Skin or additional blood symptoms	Hairy leukoplakia in AIDS
	Erythrophagocytosis

SYMPTOMS AND SIGNS

Primary Infection

The **incubation period** is reported to range from 30 to 50 days. Infection is usually asymptomatic in children, but characteristic IM develops in adolescents and young adults. In typical cases the patients have mild symptoms during the first 3–5 days such as oedema of the eyelids and meningism, especially in the evening. Later, after 7–20 days, the clinical picture is dominated by tonsillitis and general enlargement of lymph nodes, first recognized in submandibular, nuchal and axillary regions. The tonsils may be greatly enlarged, causing a variable degree of airway obstruction. Thick membranes and necrotic ulcers may be observed on the tonsils, often combined with *foetor ex ore*. Liver enzymes are usually elevated and sometimes hepatitis develops. Splenomegaly is evident in about 50% of patients. In about 5% of cases a maculopapular skin rash is seen on the body or the extremities.

In addition, ampicillin generates a rash in a high proportion of patients treated with the drug, and should therefore be avoided. Recently, antibodies to manganese superoxide dismutase (MnSOD) were found in mononucleosis patients. These autoantibodies inhibited the dismutation of superoxide

radicals; their rise and fall coincided with the clinical symptoms and they are suspected to be of pathogenetic importance.
Differential diagnosis. Initially leukopenia may be found. An increase in white cell count, especially mononuclear cells including lymphocytes and atypical mononuclear cells. The atypical cells, enlarged lymphocytes with basophilic/vacuolated cytoplasm, make up 10–20% of the leukocytes. Many of those are atypical T-lymphocytes, probably killer cells, being able to lyse EBV-containing lymphocytes, a process considered vital for the control of the disease. A cytomegalovirus infection may give similar white blood cell findings, but there is no tonsillitis or splenomegaly and no heterophile antibodies. A bacterial tonsillitis can be misdiagnosed as mononucleosis, but lacks the typical blood picture and heterophile antibodies. Minor involvement of the liver with raised transaminase levels in serum is regularly seen. Icterus sometimes appears as the first symptom, and hepatitis caused by other agents must be excluded. Mesenteric EBV adenitis can be misdiagnosed as appendicitis.

CLINICAL COURSE

The primary infection is usually asymptomatic in children. However, in adolescents and young adults mononucleosis develops. A typical disease lasts for about 10 days (1–4 weeks) with fever and sore throat. There may be a prolonged convalescence with tiredness, fatigue and low-grade fever lasting for weeks or months. A more serious course is often seen in immunodeficiencies and certain malignancies, and some may develop a chronic EBV infection or tumours.

COMPLICATIONS

Serious complications are **splenic rupture**, **severe airway obstruction**, **encephalitis** or **cardiac arrest**. Other complications are **meningoencephalitis**, **thrombocytopenia**, **haemolytic anaemia**, **haemophagocytic syndrome**, **pneumonitis** and **hepatitis**. Some patients develop a **prolonged**, **relapsing disease**, frequently associated with pneumonitis, hepatitis, or abnormal haematological findings, which can be lethal.

Chronic fatigue syndrome has been related to EBV infection. Extreme fatigue, muscle weakness, decreased memory combined with other symptoms resembling EBV infection: sore throat, low-grade fever and painful lymph nodes constitute the syndrome which lasts for at least 6 months. However, the findings of parameters indicating reactivated or active EBV infection are not routinely reported. In AIDS or AIDS-related complex EBV may be reactivated. EBV antibodies are elevated and the proportion of EBV-positive B-lymphocytes increases together with increased oral shedding of EBV. Also lymphadenopathy and hyper-IgG production in some HIV-infected persons may be caused by EBV.

Oral hairy leukoplakia in symptom-free HIV carriers indicates a poor prognosis. About 30% of HIV-infected persons get this complication which consists of 5–30 mm large skin wart-like lesions with 'hairy' surface laterally on the tongue. EBV is replicating in the upper epithelial layers.

Kawasaki's disease (mucocutaneous lymph node syndrome of children) is associated with potential lethal **coronary artery aneurysms**. A relatively high incidence of EBV DNA sequences in peripheral blood mononuclear cells indicates an association to EBV.

EBV-ASSOCIATED TUMOURS

EBV has been associated with Burkitt's lymphoma, nasopharyngeal carcinoma, Hodgkin's disease, lethal midline granuloma, X-linked lymphoproliferative syndrome and T-cell lymphoma.

Burkitt's lymphoma was discovered in East Africa, especially among children of 5–10 years of age, and especially in regions with malaria. EBV is postulated as being the aetiological agent, since:

- EBV DNA copies are found in high numbers in the tumour cells.
- EBNA antigen is expressed in nearly all tumours.
- Antibodies to EBV antigens are elevated in patients compared with matched controls.

Nasopharyngeal carcinoma is common in southern parts of China, Taiwan, Hong Kong, Singapore and Malaysia where it is the most common tumour among men. The disease is rare in other countries. Inheritance or special environmental conditions have been suggested as precipitating factors. Usually antibody levels against the early antigen of EBV (EA) and viral capsid antigen (VCA) are high, especially IgA. This has also been found in patients with other tumours: thymus, parotid, palatine tonsil and supraglottal larynx. EBV genome is found at higher frequencies in cells of **Hodgkin's lymphoma** or **lethal midline granuloma** than in non-Hodgkin's lymphoma, indicating a possible aetiological relationship.

In **X-linked lymphoproliferative syndrome** EBV develops a life-threatening disease with fatal IM and malignant lymphoma in about 75% of cases. The first patient successfully treated with allogeneic bone marrow transplantation was reported in 1994.

The numbers of **lymphomas** in HIV-infected individuals are several times higher than the number occurring in control populations. Among these, 25% are histologically Burkitt's lymphomas, where a majority showed the EBV genome. The dominating lymphoma in HIV-infected patients, however, is non-Hodgkin's lymphoma where the EBV genome is found in less than 50%. These tumours are often extranodal, sometimes intracerebral.

Immunosuppressive treatment may be complicated with the development of malignant lymphomas in 1–2% of cases. Especially children with primary EBV infection are susceptible, and may develop B-cell lymphomas with demonstrable EBV genome. The tumours often start as a polyclonal hyperplasia of B-lymphocytes that ends as monoclonal tumours. Organ transplantation has become increasingly associated with a post-transplant lymphoproliferative syndrome (PTLS), implicating up to 1% of liver and heart recipients.

THE VIRUS

The Epstein–Barr virus is a DNA virus belonging to the subfamily *Gammaherpesvirinae* of human herpesviruses. It is a 172 kb linear, double-stranded molecule with GC content of 59%. Two types can be detected by nested PCR of the EBNA-2 region of the genome. The antigens produced are:

- Early antigens that initiate, but are not dependent on, replication.
- Late antigens, VCA, and membrane antigen (MA) that are structural components of the viral particle. Antibodies to MA may neutralize the virus.
- Latent phase antigens, the EB nuclear antigen (EBNA) and the latent membrane protein (LMP), probably a component of the lymphocyte detected membrane antigen (LYDMA) that help killer cells detect lymphocytes immortalized by EBV. There are several EBNAs. EBNA-2 is required for initiation of immortalization. The HR-1 and Daudi strains of EBV are examples of viruses with deletions affecting the ability to generate immortal cells. Other strains, such as B95–8, have deletions with unknown effects.

Epstein–Barr virus enters the cell using the same receptor as the complement factor C3d. It goes directly into latency without complete replication.

B-lymphocytes that contain the EBV genome become *immortalized* or *transformed*. Upon cultivation such B-lymphocytes will demonstrate a capability to grow continuously. The EBV antigens can be demonstrated and the EBV genome by PCR or hybridization. Most of the EBV genomes of the immortalized cells are episomal, double-stranded, circular DNA. However, some may be integrated in the cellular genome, a process being enhanced by B-cell mitogens. Only one of these intracellular forms of EBV is necessary for immortalization. The process of immortalization is complex and involves a number of host cell and viral gene products.

Upon cultivation of EBV-infected B-lymphocytes, less than 10% form continuous cell lines. Immortalized lymphocytes retain their differentiation, such as immunoglobulin production. In addition they may produce virus or viral antigens.

Epithelial cells of oropharynx, parotis and uterine cervix support replication *in vivo*. However, the viral receptors disappear when cultivated. Epithelial cells of nasopharyngeal carcinoma produce virus in culture.

EPIDEMIOLOGY

EBV is found all over the world and shows no seasonal variation. IM is a moderately contagious disease with an attack rate of 10–38% among susceptible family contacts. At 5 years of age about 50% have antibodies against EBV. In children the infection is usually asymptomatic. A new wave of infection comes in the second decade, particularly so in the higher socio-economic classes, probably because fewer of them have been infected during childhood. Among adults 90–95% have antibodies against EBV.

THERAPY AND PROPHYLAXIS

Since more than 95% of the IM patients recover without specific therapy, and specific therapy does not show significant clinical benefit, the treatment is usually symptomatic. The duration of fever is reduced to 2 to 5 days using prednisolone (40 mg/day decreased to 5 mg/day by the 12th day). However, this treatment has gained no general acceptance because of the fear of inducing myocarditis or meningoencephalitis. Corticosteroids may be used in airway obstruction, severe thrombocytopenia, haemolytic anaemia or selected cases of prolonged prostration. Some also use similarly administered corticoids in involvement of the central nervous system, myocarditis, or pericarditis, starting at a higher dose (prednisone, 60–80 mg/day). Metronidazole has been claimed to reduce the pharyngeal symptoms, but neither this drug nor chloroquine, which has also been tried, showed any significant effect on general health. In randomized trials aciclovir showed no significant clinical effect. Oropharyngeal EBV replication, however, was temporarily inhibited. Phosphonoacetic acid (PAA), adenine arabinoside (ara-A), desciclovir, sorovudine (BV-ara-U) and interferon-α and -γ have demonstrated an ability to inhibit oropharyngeal shedding, but significant clinical benefit was not observed. Oral hairy leukoplakia, however, showed a significant response to aciclovir therapy. Aciclovir is also included in treatment of erythrophagocytosis, but needs to be combined with immunoglobulin and α-interferon. Cytotoxic drug therapy, sometimes combined with irradiation, is dominating therapy for EBV-related or unrelated lymphomas in immunodeficiencies, or HIV infection. Surgical removal combined with cytotoxic drugs may be required as treatment of lymphomas in transplanted patients if the tumours are monoclonal and/or show extranodal localization. However, discontinuation of the immunosuppressive treatment may be the way to stop the tumour growth, but may require removal of a transplant. *Ex vivo* generated EBV-specific CTLs may prevent or treat these lymphomas. Burkitt's lymphoma is highly sensitive to cytotoxic drugs.

LABORATORY DIAGNOSIS

The aetiological diagnosis of IM must be based on serology. A slide agglutination test for heterophile antibodies is easiest to perform. The heterophile antibodies of mononucleosis are not directed against viral antigens but will react with red blood cells from sheep, horse, goat and camel, and they have a distinct absorption pattern with other tissues of animal origin. The heterophile antibodies usually appear in the first week of the illness and may last for 8 weeks. The heterophile antibodies are pathognomonic for the disease with a sensitivity of about 95% in teenagers and adults. However, in children these antibodies may be absent. Specific serology for EBV includes detection of viral capsid antigen (VCA) and nuclear antigens EBNA-1 and EBNA-2. Acute infection is characterized by early detection of VCA antibodies. Antibodies to the nuclear antigens appear weeks to months after acute infection. For posttransplant lymphomas specific EBV serology is required.

The EBV may be detected in 80–90% of patients with IM by culturing oropharyngeal washings. However, diagnostic EBV cultures are laborious, and the interpretation is not clear since virus is excreted for weeks or months after IM. Hybridization of viral DNA and amplification (PCR) can detect EBV DNA with a maximal sensitivity of about 10 genomes. Detected genomes may also be typed using specific probes (EBV1 and EBV2). The quality of these reactions should be controlled by using cell lines infected with different EBV types, or by using plasmids containing the actual DNA fragment of EBV. A mononuclear lymphocytosis of 60–70% of total white cell count is found in the second week of IM, and peaks in the second or third week, usually with about 30% atypical lymphocytes.

ONCE SMITTEN, FOREVER RIDDEN

23. HUMAN HERPESVIRUS 6 (HHV-6)

Roseola infantum = exanthem subitum = sixth disease.

J. A. McCullers

HHV-6 is a ubiquitous virus and the most common cause of roseola, although the majority of infections are asymptomatic. The virus can remain latent and frequently reactivates in immunocompromised hosts, although potential disease associations with reactivation or primary infection in these patients are only now being elucidated.

TRANSMISSION/INCUBATION PERIOD/CLINICAL FEATURES

HHV-6 is transmitted by direct contact with infectious particles from the saliva or genital secretions of an infected individual, typically during close contact between parent and child in the first year of life. The virus remains latent in the host and can be shed asymptomatically at intervals throughout life or can reactivate during periods of immunocompromise.

SYMPTOMS AND SIGNS

Systemic:	High Fever, Morbilliform Rash
Local:	None (During Typical Infection)
Other:	*See* 'Complications'

Typical primary illness occurs in children 6 months to 2 years of age and is characterized by high fever lasting 3–5 days, followed by an abrupt defervescence and appearance of a morbilliform rash which may last 2–24 hours.

COMPLICATIONS

Febrile seizures complicate a significant percentage of HHV-6 infections in small children. Hepatitis, bone marrow suppression, pneumonia, encephalitis and infection with CMV have been associated with HHV-6, primarily during reactivation in immunosuppressed hosts such as transplant patients.

THERAPY AND PROPHYLAXIS

Specific therapy is not available for roseola, although supportive care such as hospitalization of infants for dehydration is sometimes necessary. Ganciclovir may be useful in immunosuppressed hosts.

LABORATORY DIAGNOSIS

Diagnosis of roseola is clinical. Laboratory diagnosis in immunosuppressed patients by seroconversion or qualitative PCR is available commercially but is rarely helpful clinically.

CLINICAL FEATURES

SYMPTOMS AND SIGNS

The cardinal sign of infection is fever. The range of fever experienced with infection is wide, but temperatures are frequently high (>40°C). In classic roseola fever persists for 3–5 days followed by an abrupt defervescence. Shortly after defervescence a pink, non-pruritic maculopapular rash appears. The exanthem typically starts on the neck or back, sparing the face, and progresses down over the trunk and extremities. It persists for 2–24 hours, then fades quickly and can often be missed on examination, occurring while the child is sleeping. Although classic descriptions of roseola infection include this rash, it is now recognized that the majority of cases of primary HHV-6 infection are either asymptomatic or consist of fever alone without rash. Febrile seizures at the height of fever are a common complication. Supportive care is generally recommended with dehydration or the need to rule out more serious illnesses in infants being the most common reasons for hospitalization. Cases in immunocompetent adults appear to be characterized by fever without rash when overt symptoms are seen. Fever or bone marrow suppression are commonly seen during reactivation in immunosuppressed hosts, particularly patients undergoing bone marrow or solid organ transplant.

Differential diagnosis. Classic roseola can also be caused by human herpesvirus-7 (HHV-7), and occasionally by other viruses. The findings and clinical course are the same, although roseola from HHV-7 may appear at a slightly older age. Fever and/or rash in an infant can be caused by a number of viruses, including enteroviruses, coxsackievirus, echovirus, measles and rubella, as well as streptococcal or staphylococcal bacterial infections.

CLINICAL COURSE

The mean incubation period appears to be 9 to 10 days. Symptoms from classic roseola typically end with defervescence and appearance of the morbilliform rash. Febrile seizures frequently occur at the height of fever and are self-limited. The clinical course of less well-defined complications is not well characterized.

COMPLICATIONS

Febrile seizures are the most common complication of HHV-6 infection. Infection with this virus may account for one-third of all febrile seizures in infants. Generally these are simple in nature and do not recur. A number of complications have been associated with reactivation of HHV-6 in immunosuppressed hosts, although it remains uncertain in many cases whether HHV-6 itself, HHV-6-induced reactivation of other infectious agents such as CMV, or some other event for which HHV-6 reactivation is merely a marker is responsible for the observed disease manifestations. Fever, bone marrow suppression, hepatitis, pneumonitis, encephalitis, graft or organ rejection, and an increase in infections from CMV and fungi have been associated with HHV-6.

THE VIRUS

HHV-6 is the sixth of eight described human herpesviruses. Like the other members of the group, it has an electron-dense core composed of double-stranded DNA (160–170 kbp long) surrounded by nucleocapsid in the shape of an icosahedron. A tegument surrounds the capsid, and a trilaminar membrane surrounds the entire structure. The nucleocapsid is 90–110 nm in diameter, and the mature virion ranges from 160 to 200 nm in diameter. The genome of HHV-6 is most closely related to HHV-7; of all other herpesviruses these two similar viruses are most closely related to CMV. Two variants of the virus have been described, HHV-6A and HHV-6B. Although both utilize the ubiquitous human CD46 as a receptor, there are likely to be other factors involved in attachment or fusion as the two variants display different cellular tropisms in tissue culture and in the host. These tropisms may relate to observed disease associations. HHV-6B causes nearly all HHV-6-associated roseola and has been implicated in most of the observed complications in immunosupressed hosts, while HHV-6A has only rarely been implicated in disease. HHV-6B grows well in mature lymphocytes while HHV-6A replicates better in immature lymphocytes.

EPIDEMIOLOGY

HHV-6 is a ubiquitous virus which is found worldwide and appears to infect most children by the age of two years. The virus can be found in the saliva of around 70% of asymptomatic adults, closely matching the seroprevalence in the adult population. Transmission is thought to occur by direct contact with infectious secretions passing from asymptomatic adults to seronegative children, typically from mother to child. Virus can be recovered from saliva, genital secretions and brain at autopsy in asymptomatic adults. Virus is generally only recovered from blood during viraemia such as at the height of fever during primary infection (roseola) or during reactivation in immuno-compromised hosts, but can be detected in latent form with sensitive techniques such as PCR.

THERAPY AND PROPHYLAXIS

There is no specific antiviral therapy for roseola. Supportive care such as hospitalization of infants for dehydration is sometimes necessary. Ganciclovir has antiviral activity *in vitro*, and it has been suggested that severe illness associated with reactivation in immunosuppressed hosts be treated with ganciclovir or foscarnet. Clinical efficacy data are not available.

Prophylaxis is not available.

LABORATORY DIAGNOSIS

Diagnosis of roseola is clinical. HHV-6 can be detected by culture of peripheral blood mononuclear cells during primary infection or reactivation but is rarely used due to its difficulty. Laboratory diagnosis in immunosuppressed patients by seroconversion or qualitative PCR is available commercially but is rarely helpful clinically due to the high rate of seropositivity (approaching 100% at age 2 years) and frequency of positive PCR in sites such as blood, saliva and brain tissue. Quantitative and real-time PCR techniques are currently found only in research laboratories and may prove more useful for delineating disease associations by following serial values from blood.

PREVENTING TRAVEL SICKNESS?

24. HEPATITIS A VIRUS

Hepatitis A has been called infectious hepatitis, epidemic hepatitis and short incubation hepatitis.

M. Degré

Hepatitis A is a viral infection resulting in an acute necroinflammatory disease of the liver. The clinical features are age-dependent, mostly mild or sub-clinical in children, while in adults dominated by jaundice and general fatigue.

TRANSMISSION/INCUBATION PERIOD/CLINICAL FEATURES

Virus is excreted in the faeces mostly during the late phase of incubation period and to a lesser extent during the early phase of the disease. It is transmitted via the faecal–oral route, through contaminated water and food and direct contact, especially under poor hygienic conditions. The incubation period is approximately 4 weeks (2–6 weeks) and dependent on the size of inoculum.

SYMPTOMS AND SIGNS

Prodromal phase:	Malaise, Weakness, Mild Fever, Nausea and Vomiting, Anorexia, Headache, Abdominal Discomfort
Icteric phase:	Jaundice, Itching, Dark Urine, Pale Stools

The majority of young patients are symptomless or they have mild general symptoms. The prodromal phase that lasts 2–7 days is followed by the icteric phase lasting usually 1–2 weeks, often longer in adults. Some patients show general fatigue for several weeks, or even months. Extrahepatic manifestations are uncommon.

COMPLICATIONS

Fulminant hepatitis occurs in less than 1% of patients. Relapsing hepatitis A and cholestatic forms have been described but are also uncommon.

THERAPY AND PROPHYLAXIS

There is no specific causal therapy available. Normal immunoglobulin has a protective effect when given before or shortly after exposure. Both inactivated and attenuated vaccines have been developed and are available.

LABORATORY DIAGNOSIS

The clinical virological diagnosis is made by demonstrating the presence of IgM antibodies in the early phase of clinical disease or by demonstrating seroconversion. Early infection can be diagnosed by the presence of IgG antibodies.

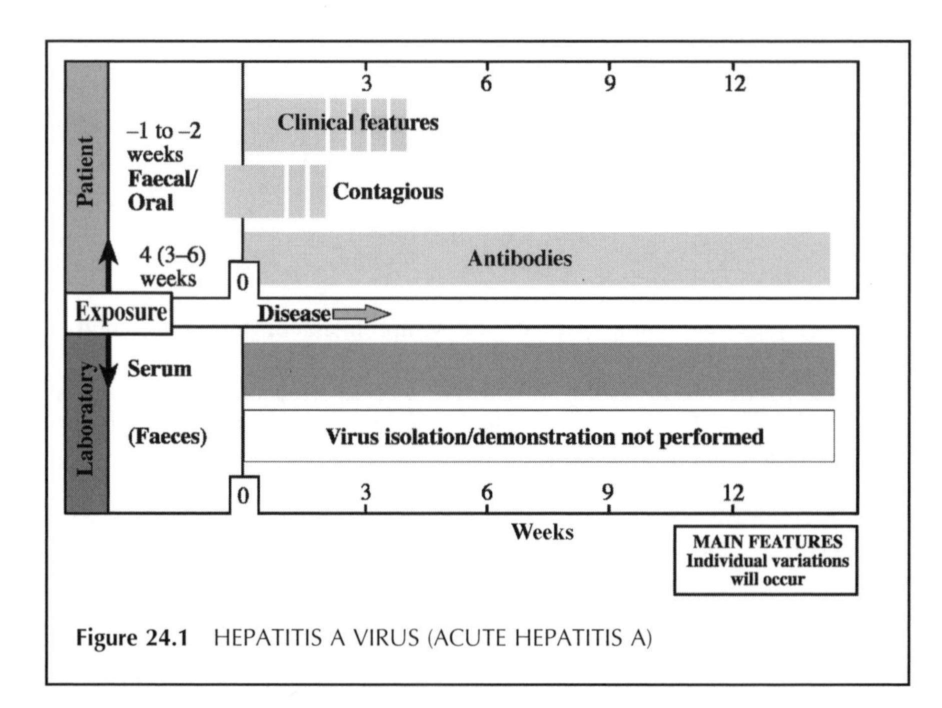

Figure 24.1 HEPATITIS A VIRUS (ACUTE HEPATITIS A)

CLINICAL FEATURES

SYMPTOMS AND SIGNS

The incubation period is usually 4 weeks (2–6 weeks). The site of the primary infection is in the alimentary tract, although the sequence of events that begins with entry via the gastrointestinal tract and eventually results in hepatitis is not well understood. A short prodromal or preicteric phase, varying from 2 to 7 days, usually precedes the onset of jaundice. The most prominent symptoms in this phase are fever, headache, muscular and abdominal pain, anorexia, nausea, vomiting and sometimes arthralgia. Hepatomegaly and leukopenia are often present during this period. In typical cases the urine becomes dark, and the stools pale before appearance of yellow discoloration of the mucous membranes and appearance of jaundice about 10 days after onset of the general symptoms. Fever and most of the general symptoms usually subside within a few days of jaundice, but in severe cases both general and abdominal symptoms may become further aggravated at this phase. Jaundice is often accompanied by itching and sometimes by urticarial or papular rashes. Liver is usually enlarged and liver function tests are abnormal with highly elevated levels of serum alanine aminotransaminase (ALT) and serum aspartate aminotransaminase (AST).

Differential diagnosis. It is not possible to differentiate between hepatitis A and acute hepatitis caused by other infectious agents, e.g. hepatitis B, hepatitis C, hepatitis E, cytomegalovirus and EBV, by clinical examination or by liver biopsy. The history and epidemiological details are of great importance in reaching a presumptive diagnosis. The precise aetiology must be confirmed by laboratory tests.

CLINICAL COURSE

Most infections run an asymptomatic course, especially in children. The disease is usually mild and of short duration in children and young adults. Clinical symptoms are likely to disappear after 10–14 days. In adults the symptoms are often more severe and long-lasting, e.g. 4–5 weeks. The liver function tests rapidly return to normal when clinical symptoms disappear, but in some cases symptoms may persist for several months, during which time the patient feels tired and often depressed. Hepatitis A virus does not cause chronic hepatitis, although long-lasting excretion of virus has been reported.

COMPLICATIONS

Fulminant hepatitis with a clinical picture of acute yellow atrophy is an uncommon but serious (lethality 0.3%) complication with high mortality. Extrahepatic complications like myocarditis and arthritis are also rare.

THE VIRUS

Hepatitis A virus (HAV) is a member of the picornavirus family. It consists of a naked icosahedral particle of 27 nm diameter. The genome is a single-stranded RNA, linear, positive-sense, 7.48 kbp, MW 2.25 million daltons. It codes for a polyprotein, which is cleaved into four major structural proteins VP1–4. The surface proteins VP1–3 are major antibody-binding sites. It was first provisionally classified as enterovirus 72, but subsequently it has been shown that both nucleotide and aminoacid sequences are dissimilar from the other enteroviruses, and it is now classified as its own genus, hepatovirus. Although minor strain variations occur, there is only one serological type, but we can differentiate between four genotypes. Hepatitis A virus is stable to treatment with organic solvents, ether and acid and is more heat-resistant than other picornaviruses; it withstands 60 °C for 1 hour. The virus is difficult to adopt to cell cultures and replicates very slowly, usually without cytopathogenic effect. *In vivo* HAV primarily multiplies in Peyer's patches in the intestinal tract and later in hepatocytes and Kupffer cells. Viral antigen can be demonstrated by immunofluorescence in liver biopsies. The liver damage is probably partly a direct result of the viral cytopathogenic effect, but T-cell-mediated immunological mechanisms seem to be important. HAV can be transmitted to chimpanzees and marmoset monkeys.

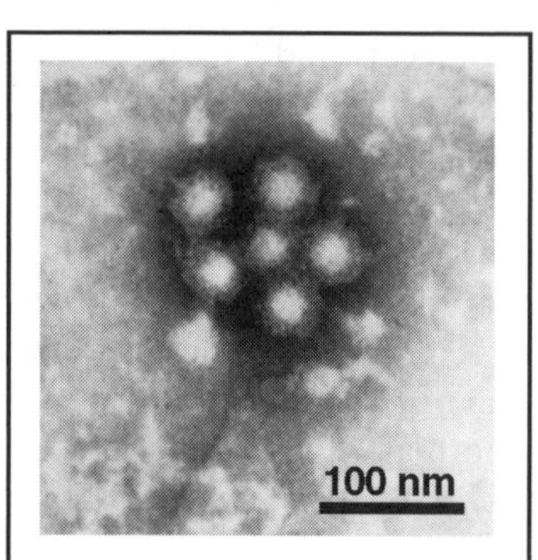

Figure 24.2 HEPATITIS A VIRUS DECORATED WITH ANTIBODY. Bar, 100 nm (Electron micrograph courtesy of E. Kjeldsberg)

EPIDEMIOLOGY

Hepatitis A has been known since the time of Hippocrates and has a worldwide distribution. The epidemiology of the disease is a function of its principal route of spread, faecal–oral transmission. Water- and food-borne epidemics are well documented. Man is the only significant reservoir, and infection provides lifelong immunity. Three major patterns of infection are known which reflect different epidemiological situations. These are demonstrated by different patterns of the age-specific prevalence of antibodies to HAV which reflect standards of hygiene and sanitation, the degree of crowding of the population and opportunities for the virus to survive and spread. In developing countries hepatitis A is endemic, and more than 90% of the adult population is immune. In industrialized countries most people are susceptible to infection and travellers to developing countries thus carry the risk of contracting the infection. The occurrence of hepatitis A in north-west European countries and in North America has been much reduced during the last decades. Antibodies to HAV have been found in 80–90% of individuals born before World War II, but only 5–20%, or even less, of those under 20 years of age.

THERAPY AND PROPHYLAXIS

There is no specific **therapy**. Bed rest is traditional in the treatment of hepatitis A and is recommended if the general situation indicates so, especially in elderly patients and during pregnancy. The preventive effect of **immunoglobulin** is well documented. Administration of 0.02–0.06 ml/kg (2–5 ml) normal immunoglobulin before exposure gives 80–90% protection for a period of 4–6 months. Even after infection, given during the early part of the incubation period, immunoglobulin can prevent clinical disease, although virus may be present in the intestines. Immunoglobulin is recommended to people travelling to endemic areas. A formalin-inactivated **vaccine** is now generally available. After two doses of vaccine almost 100% of individuals develop antibodies and are protected against infection. An attenuated vaccine has also been introduced. Spread of virus infection is to a large extent a function of socioeconomic standards and can be prevented by means of good hygiene. Infectivity of virus is destroyed by boiling for 15 minutes or at 60 °C for 30 minutes. Chlorine derivatives, formaldehyde and glutaraldehyde are all effective in routinely employed concentrations, while phenols, ether and other organic solvents are not. Isolation of the patients is not necessary as infectivity in the icteric phase is usually insignificant.

LABORATORY DIAGNOSIS

HAV is present in stools before the onset of clinical symptoms and can be demonstrated by electron microscopy. Isolation in cell cultures is difficult and not practical for diagnostic use. Detection of nucleic acid with PCR can be useful both as an epidemiological tool and in environmental studies but it is not indicated for clinical use. Antibodies can be demonstrated from the onset of the clinical symptoms. Diagnosis of acute infection requires demonstration of anti-HIV IgM antibodies or seroconversion. IgM antibodies disappear about 3–6 months after the onset of disease. IgG antibodies on the other hand persist for life and indicate immunity against reinfection. Antibodies are demonstrated by means of EIA or RIA.

RISKY BUSINESS

25. HEPATITIS B VIRUS

Previously called serum hepatitis, post-transfusion hepatitis and inoculation hepatitis.

G. L. Davis

Infection with the hepatitis B virus (HBV) can result in acute hepatitis, fulminant hepatitis, a chronic asymptomatic carrier state, chronic hepatitis, cirrhosis or hepatocellular carcinoma.

TRANSMISSION/INCUBATION PERIOD/CLINICAL FEATURES

HBV is present in blood and bodily secretions. The virus is most commonly spread by sexual contact, but it can also be spread from mother to child at birth, through contaminated needles and transfusion (extremely rare). The incubation period averages 75 days (range 1–6 months).

SYMPTOMS AND SIGNS

Systemic:	Fever, Fatigue, Malaise, Dyspepsia, Rash, Arthralgia
Local:	Hepatomegaly, Pale Stools, Dark Urine, Jaundice

Acute infection is usually mild and anicteric; only about a third of patients are aware of the infection. Complete recovery occurs in more than 95% of adults, but is unusual ($< 10\%$) if infection occurs in infancy when most infections become chronic. Chronic HBV infection may present in one of two ways. Individuals infected early in life are tolerant of the virus and often have an asymptomatic carrier state with normal liver tests. Those infected later in life usually present with chronic hepatitis and elevated liver enzymes. The latter are more likely to have symptoms and develop progressive liver damage at a faster rate.

COMPLICATIONS

About one in six patients with acute hepatitis has serum sickness-like symptoms of rash, fever and arthralgia. Fulminant hepatitis is unusual (assumed to be 1 in 200). Chronic hepatitis leads to cirrhosis in more than 20% of cases, and the risk of hepatocellular carcinoma is increased about 50-fold. Polyarteritis nodosa, glomerulonephritis and papular acrodermatitis (only in children) occur rarely in those with chronic hepatitis.

THERAPY AND PROPHYLAXIS

Acute infection can be prevented by the HBV vaccine. Infants born to HBV-infected mothers should also receive HBV immunoglobulin (HBIG), which is also recommended for postexposure prophylaxis. Interferon eliminates viral replication in 40–50% of treated patients mostly. Response is usually permanent. Lamivudine and some new nucleoside analogues may be equally effective, though viral resistance may emerge.

LABORATORY DIAGNOSIS

In an acute infection the detection of HBsAg and IgM antibody to the nucleocapsid (HBc) is characteristic, followed by development of convalescent anti-HBs antibodies. Chronic infection is indicated by the presence of HBsAg and absence of IgM anti-HBc. Viral replication occurs during the initial high replicative phase of infection and these patients are HBeAg and HBV-DNA (by a non-PCR method) positive. About 10% of patients annually will spontaneously develop into a low replication stage; becoming HBeAg and HBV-DNA (by a non-PCR method) negative and anti-HBe positive. Loss of HBsAg is extremely unusual in patients with chronic infection unless they have been treated with interferon.

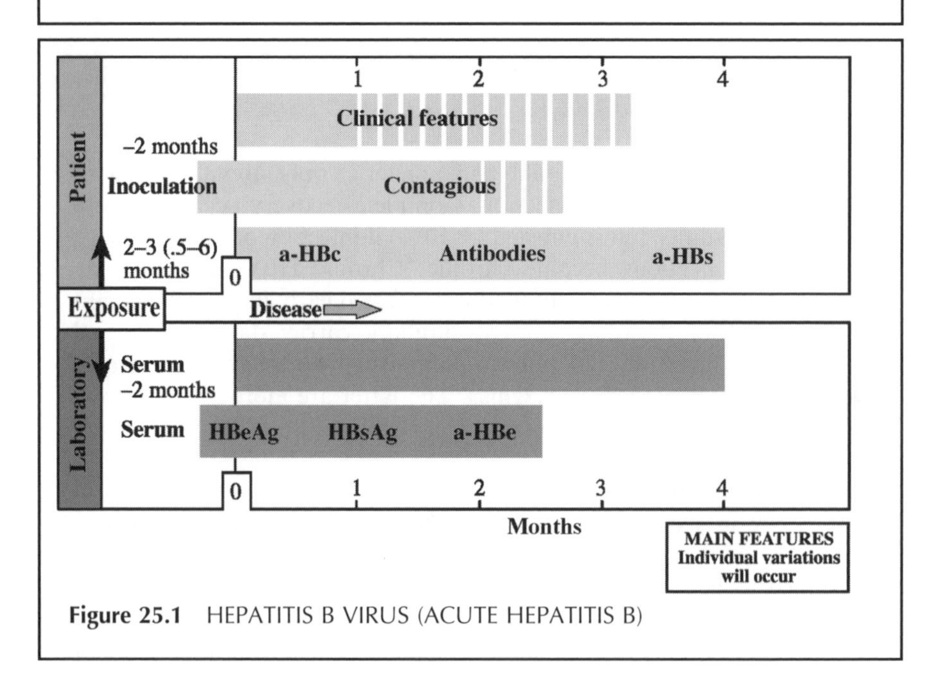

Figure 25.1 HEPATITIS B VIRUS (ACUTE HEPATITIS B)

CLINICAL FEATURES

SYMPTOMS AND SIGNS

The **incubation period** of HBV infection averages 75 days (1–6 months). Most patients have minimal or no symptoms during infection. However, 15% present with fever, arthralgia and rash, and another 15–20% present with non-specific symptoms of malaise, anorexia and nausea. About a quarter of patients will become icteric. Findings on physical examination are usually non-specific though hepatomegaly may be noted in some cases.

Differential diagnosis. The symptoms of acute hepatitis are non-specific. Similar symptoms may be seen in other forms of viral or drug-induced hepatitis. In children and adolescents, EBV should be considered. In older adults, cholecystitis, cholangitis and common bile duct obstruction become considerations. The hepatocellular pattern of liver test abnormalities (elevated AST and ALT with slight or no elevation of alkaline phosphatase) should lead to a diagnosis rather than biliary disease in most cases. An ALT level higher than 1000 U/litre is seen only in viral hepatitis, drug-induced hepatitis and ischaemic hepatic injury. Serological diagnosis is required to confirm the HBV infection.

CLINICAL COURSE

Serological markers of infection (HBsAg) appear in the blood early in the incubation period. This is followed within a few weeks by evidence of viral replication (HBeAg and HBV-DNA (by a non-PCR method)). These markers and liver enzymes reach their highest level at about the time of onset of symptoms, which abate within 2–3 weeks but may persist for months. In the majority of patients (>90%), HBsAg disappears from the blood as the liver enzymes normalize. Convalescent antibody (anti-HBs) will develop in most of these patients. Persistence of HBsAg and HBeAg for more than 10 weeks indicates that chronic infection is likely to evolve. Chronic hepatitis persists for years. The initial phase is characterized by high levels of viral replication with detectable HBeAg and HBV-DNA (by a non-PCR method). Patients are infectious and at risk for progressive liver injury during this period. Cirrhosis develops in 20–50% of cases. About 10% of patients each year will have a spontaneous reduction in viral replication in which HBeAg and HBV-DNA (by a non-PCR method) become undetectable. These individuals are commonly no longer infectious and their liver disease becomes inactive. However, it is very unusual for patients with chronic HBV infection to clear the infection and lose HBsAg unless they receive antiviral treatment.

COMPLICATIONS

The major long-term risks of chronic HBV infection are cirrhosis with hepatic failure and hepatocellular carcinoma. Cirrhosis develops in 20–50% of cases and is associated with a poor prognosis, particularly if HBeAg is present. Between 25 and 50% of cirrhotic patients will develop liver failure. Hepatocellular carcinoma occurs most frequently in cirrhotic patients who have had long-standing liver disease. The risk is increased >50-fold compared with the normal population. Extrahepatic manifestation of infection such as polyarteritis nodosa and glomerulonephritis are rare, but when present are often more problematic than the hepatitis. An increasingly common problem is the development of viral variants. The most common is the precore mutant which presents with active replication (HBV-DNA (by a non-PCR method) positive) but no HBeAg. The patients behave like others with chronic hepatitis, but their course may be somewhat more aggressive.

THE VIRUS

HBV is a member of the *Hepadnaviridae* family (Figure 25.2). It is the only DNA virus among the agents which commonly cause viral hepatitis. The viral particle (called the Dane particle) is 42 nm in diameter. The lipoprotein (HBsAg) which encoats the virus is seen not only as a viral envelope but also by electron microscopy as free non-infectious tubular and spherical structures. These forms of HBsAg circulate in considerable excess compared with the virion and may play a permissive role in viral persistence. However, HBsAg may be present in blood when replication cannot be documented. Therefore, the presence of HBsAG does not necessarily imply contagiousness. HBV replicates in hepatocytes and possibly in peripheral blood mononuclear cells. Its genome is the smallest of all known animal DNA viruses. The replicative process is unusual in several aspects; it has an efficient genomic design of four overlapping open reading frames, it utilizes successive strand synthesis and reverse transcription similar to retroviruses, and it has both glucocorticoid- and hepatocyte-specific enhancing elements. Viral replication can be documented by measuring HBeAg (a component of the core gene product) or HBV-DNA (by a non-PCR method) in serum.

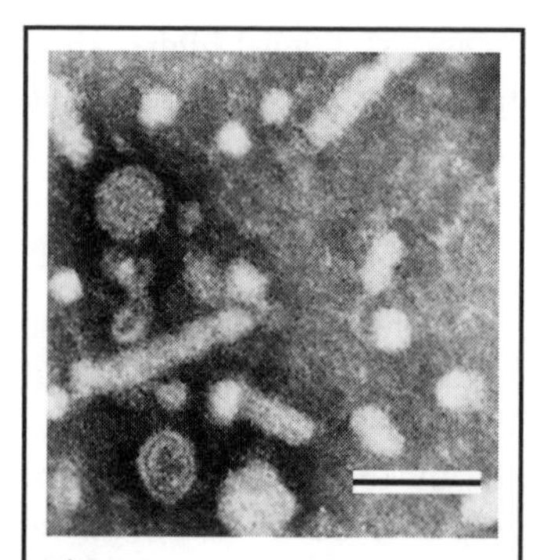

Figure 25.2 HEPATITIS B VIRUS AND HBsAg PARTICLES. Bar, 100 nm (Electron micrograph courtesy of E. Kjeldsberg)

EPIDEMIOLOGY

HBV infection is a formidable immense worldwide problem. More than 200 million people are chronically infected. The prevalence is highly variable in the Far East, and in Mediterranean and Eastern European countries, whereas in sub-Saharan Africa the endemic rates are highest, with as many as 20% of the population being infected. In North America and Western Europe the infection is not common (0.1–0.2%). The major route of infection in high endemic areas is perinatal. In countries of low endemicity, the major routes of infection are sexual and shared needles amongst intravenous drug users. The latter group is notoriously difficult to target by vaccination. However, universal or extensive vaccination may be the only practical means of achieving a significant reduction of HBV prevalence.

THERAPY AND PROPHYLAXIS

HBV infection is a preventable disease. **Vaccination** has been available since the early 1980s, but compliance has been poor in low endemicity areas and the high cost has delayed widespread usage in high endemicity areas. Postexposure prophylaxis with high titre **hepatitis B immunoglobulin (HBIG)** provides short-term passive protection but is only about 75% effective. No specific **treatment** is required for acute hepatitis B since most individuals will clear the infection spontaneously. Patients with chronic hepatitis should be evaluated for treatment with **alpha interferon**. Overall the likelihood of clearing the infection with interferon is 40–50%, and for patients with elevated ALT and a low quantity of HBV-DNA (by a non-PCR method) the response is significant and usually permanent. Many patients will even lose HBsAg over a period of 4–5 years after responding to interferon. HBV is quite sensitive to several new nucleoside analogies, but the effect of these drugs is usually transient unless used for at least 12 months. Drug resistant variants may appear, particularly after prolonged therapy.

LABORATORY DIAGNOSIS

HBsAg is the most important serological marker for identifying infection. It is present early in acute infection, disappears with resolution of infection and persists in chronic infection. IgM anti-HBc is essential for the diagnosis of acute infection, but is also seen occasionally in very active chronic hepatitis. Anti-HBc antibodies develop and persist after all HBV infections. The loss of HBsAg and development of anti-HBc signals resolution of acute infection. Anti-HBs also occurs post vaccination, but anti-HBc will not be present in such cases. Chronic infection is manifested by persistent HBsAg. Markers of viral replication such as HBeAg and HBV-DNA (non-PCR method) are detectable during the early high replication phase, but are not detectable during the later quiescent low replication phase. HBeAg is not a reliable marker of HBV replication when a precore variant is responsible for the infection. Such cases will be HBeAg negative, anti-HBe positive, but HBV-DNA (by a non-PCR method) positive.

FINALLY NAMED, SEE?!

26. HEPATITIS C VIRUS

G. L. Davis

Infection with the hepatitis C virus (HCV) can result in acute hepatitis, a chronic asymptomatic carrier state, chronic hepatitis, cirrhosis or hepatocellular carcinoma.

TRANSMISSION/INCUBATION PERIOD/CLINICAL FEATURES

HCV is present in blood. Although hepatitis C was the major factor responsible for the large number of post-transfusion hepatitis cases in the past, this risk has nearly been eliminated by screening of blood donors. Today, the major known route of spread of this infection is sharing of needles among intravenous drug abusers. The incubation time is approximately 7 weeks.

SYMPTOMS AND SIGNS

Systemic:	Malaise, Nausea, Anorexia
Local:	Hepatomegaly, Splenomegaly, Cholestasis, Jaundice

Acute infection is usually mild and anicteric; only about a quarter of patients are aware of the infection. Complete recovery is unusual; more than 55–85%, depending upon the age of acquisition, develop chronic hepatitis with elevated liver tests and an even higher proportion have persistent viraemia. Chronic HCV infection may present in one of two ways, either a 'carrier state' with normal liver enzyme levels or typical chronic hepatitis. Persons with HCV infection and normal liver enzymes are usually asymptomatic, but most have some degree of liver injury seen on liver biopsy. Patients with chronic hepatitis typically have elevated liver tests, although these may be only slightly increased. Symptoms are present in only half of patients. It is probably safe to say that most of the estimated 175 million persons with this infection are not aware of it.

COMPLICATIONS

Chronic hepatitis leads to cirrhosis in more than 20% of cases. The risk of hepatocellular carcinoma is increased. HCV is the major cause of essential cryoglobulinaemia. This occurs in about half of cirrhotic patients and can be associated with rash, purpura, vasculitis, arthralgia and glomerulonephritis.

THERAPY AND PROPHYLAXIS

There is no vaccine or immunoglobulin to prevent HCV infection. The only agents with known activity are type I interferons (alpha and beta). More than half of patients permanently eradicate virus when treated with pegylated (long-acting) interferon and ribavirin for 6 to 12 months. The response and required duration of treatment is highly dependent on viral genotype.

LABORATORY DIAGNOSIS

Serological diagnosis of acute infection can be troublesome. Antibody to HCV is present in only about two-thirds of patients at presentation, so serial samples may need to be tested in order to confirm the diagnosis. Chronic infection is diagnosed in patients with elevated liver enzyme levels and anti-HCV antibodies. Individuals with anti-HCV antibodies and normal liver enzyme values could be falsely positive. The test can be confirmed using radioimmunoblot assay (RIBA) or detection of HCV-RNA by an amplified method such as PCR or TMA test.

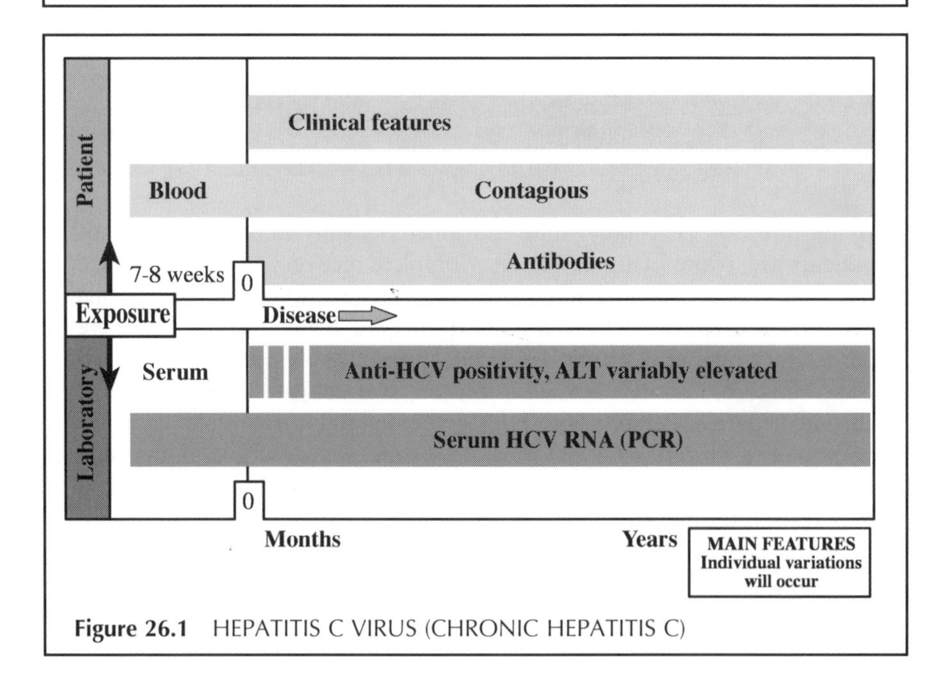

Figure 26.1 HEPATITIS C VIRUS (CHRONIC HEPATITIS C)

CLINICAL FEATURES

SYMPTOMS AND SIGNS

The **incubation period** of HCV infection averages 50 days (14 days to several months). Most patients have minimal or no symptoms during acute infection. As a result, it is quite unusual for infection to be recognized during this stage. Findings on physical examination are usually non-specific though hepatomegaly may be noted in some cases.

Differential diagnosis. As with all forms of acute viral hepatitis, the symptoms are non-specific. Similar symptoms may be seen in other forms of viral or drug-induced hepatitis. Careful questioning of the patient will identify risk factors for HCV infection in about two-thirds of cases. However, since seroconversion is often delayed, a careful search for other causes of liver disease is important. Evidence for infection with EBV or HHV-6 should be sought. Particularly in adults, cholecystitis, cholangitis and common bile duct obstruction should be considered. The hepatocellular pattern of liver test abnormalities (elevated AST and ALT with slight or no elevation of alkaline phosphatase) should lead to a diagnosis of hepatitis rather than biliary disease in most cases. An ALT level higher than 1000 U/litre is seen only in viral hepatitis, drug-induced hepatitis and ischaemic hepatic injury.

CLINICAL COURSE

Acute infection resolves in only 10–35% of cases. About 55–80% of patients will develop chronic hepatitis with persistent elevation of the liver enzymes. Another 10% will develop persistent viraemia with normal liver tests. Spontaneous resolution of the infection beyond this stage is extremely unusual. Chronic hepatitis C is usually minimally symptomatic and only slowly progressive, usually over two to three decades. Thus, most patients do not come to medical attention until long after the onset of their infection. However, the disease will progress slowly in the majority of patients and cirrhosis develops in about 20% of cases. HCV infection is now the leading cause of hepatocellular carcinoma in some parts of the world, e.g. Japan. Liver failure results in about a quarter of cirrhotic patients. Chronic hepatitis C is the leading indication for liver transplantation in the USA.

COMPLICATIONS

The major long-term risks of chronic HCV infection are cirrhosis with hepatic failure and hepatocellular carcinoma. Cirrhosis develops in 20% of cases and has a poor prognosis. About 25% of cirrhotic patients will develop liver failure. About 1 in every 20–30 infections will die from complications of the liver disease caused by the infection. Hepatocellular carcinoma occurs most frequently in cirrhotic patients who have longstanding liver disease. HCV is now the most common risk factor for hepatocellular carcinoma in many areas of the world, including Japan.

Extrahepatic manifestations of infection include mixed essential cryoglobulinaemia and glomerulonephritis. Although cryoglobulinaemia is common (>50% of patients with cirrhosis) it is not often symptomatic or progressive.

THE VIRUS

HCV is an RNA virus of the hepacivirus genus of the Flaviviridae and is related to viruses of the animal Pestivirus genus. The physical characterization of HCV is preliminary since the viral particle has not been visualized with certainty. It has a lipid membrane since it is inactivated by chloroform and is probably 30–45 nm in size. HCV replicates in hepatocytes and possible in peripheral blood mononuclear cells. Its genome is similar in size and structure to other flaviviruses. Viral replication can be documented by HCV-RNA RT-PCR. HCV is extremely heterogeneous and, like many RNA viruses, has a high mutation rate. These genotypes are responsible for the various manifestations of the disease worldwide, since they are geographically distributed and have different epidemiology, natural history and response to antiviral treatment.

EPIDEMIOLOGY

HCV infects approximately 1.5–2% of the world's population, in particular in Japan, Africa, the Mediterranean countries and the Middle East. There are 3.5 million infected in the USA, making this the most common form of liver disease. The prevalence and natural history of the infection varies from country to country depending on the routes of infection and the common viral genotypes. The major route of infection was previously blood transfusion. Before screening of donors, the risk of post-transfusion hepatitis was 10–30% of recipients. The risk with extensive screening of donors today is about 0.03% per unit. The major genotypes with transfusion-acquired infection are 1a and 1b. The major routes of infection now are shared needles among intravenous drug abusers, exposure among health care workers, and perhaps also sexual. In about 15–30% of cases a risk factor cannot be identified. In the USA and Europe the genotypes 2 and 3 are more common among drug abusers than in the general population.

THERAPY AND PROPHYLAXIS

HCV infection can only be prevented by avoiding contact with the virus. There is no **vaccine** and the heterogeneity of the virus makes it difficult to develop a conventional vaccine in the near future. Pre- and postexposure prophylaxis with **immunoglobulin** is ineffective. Interferon (alpha or beta) is the only treatment known with activity against HCV. Interferon treatment reduces the chronicity rate in acute infection. Unfortunately most patients are asymptomatic during this phase and therefore do not come to medical attention. The current standard of care for chronic hepatitis C infection is pegylated interferon in combination with oral ribavirin. A 12 month course of this drug combination is effective in permanently eradicating infection in about 55% of cases. Viral genotypes 2 and 3 respond better and 80% clear infection with just 6 months of treatment.

LABORATORY DIAGNOSIS

Anti-HCV is the most important serological marker for identifying infection. It may not be detectable early in acute infection, but will develop in later serum samples. Anti-HCV is almost always present during chronic infection. In the presence of elevated liver tests, anti-HCV is highly specific and need not be confirmed by immunoblotting or HCV-RNA testing. However, in patients with normal liver enzymes, anti-HCV should be confirmed by one of these techniques. HCV-RNA determinations are usually not helpful for diagnosis other than in carriers with normal liver tests. They may be used, though, in following the response to treatment and modifying therapeutic regimens. Viral genotyping is currently of limited value in the clinic. It is possible that future treatment will be modified according to viral genotype, HCV-RNA levels and the degree of hepatic injury.

. . . OBSERVED LEAVING IN DISGUISE . . .

27. HEPATITIS D VIRUS

Previously known as delta agent or delta virus.

G. L. Davis

The delta agent, now designated hepatitis D virus (HDV), multiples only in persons harbouring hepatitis B virus, after co-infection or HDV superinfection.

TRANSMISSION/INCUBATION PERIOD

HDV infection occurs only in HBV-infected patients. Not surprisingly, the epidemiology of HDV is quite similar to HBV. Most infected patients have acquired the infection sexually or parenterally by sharing of needles. Infection can occur concurrently with acquisition of HBV (co-infection) or subsequent to HBV infection in a patient with chronic hepatitis B (superinfection). Because of its dependence upon the presence of hepatitis B infection, the incubation period has not been precisely defined.

THE VIRUS

HDV is a unique virus which consists of a single-stranded circular RNA and the delta antigen (HDAg) encoated by the lipoprotein coat of the hepatitis B virus (HBsAg). Hepatitis D virus particles are 35–37 nm in diameter.

CLINICAL FEATURES AND COURSE/COMPLICATION

Co-infection of HDV and HBV is usually self-limited. Since chronic infection of HBV would be required to perpetuate HDV, chronic HDV infection results in only about 2% of acute cases. The severity of co-infection varies, but in most cases is not different from acute hepatitis B. Co-infection should be suspected when the acute hepatitis is severe, when a bimodal peak of raised liver enzyme levels occurs, or in the presence of fulminant hepatitis.

Superinfection with HDV occurs in patients with chronic HBV infection, usually resulting in chronic HDV with delta antigen persisting in the liver. Superinfection should be suspected when previously stable chronic hepatitis B suddenly worsens. Diagnosis is easier than in acute co-infection since both IgM and IgG anti-HDV quickly become detectable in serum.

THERAPY AND PROPHYLAXIS

Treatment options for chronic hepatitis D are limited. Although α-inferferon is beneficial, it requires high doses for long periods of time and relapse is common when the therapy is stopped. Hepatitis B vaccine and passive immunoprophylaxis prevent HBV infection and thereby avert acute co-infection with HDV. However, other than avoiding risk-associated behaviour, there is no specific prophylaxis to avoid superinfection.

LABORATORY DIAGNOSIS

Direct confirmation of HDV infection is made by demonstrating HDAg or HDV-RNA in serum or liver, but these are generally research tools and not widely available. Delta antigen may be detected in blood during the first weeks of infection, and later antibodies against this antigen can be demonstrated by RIA or ELISA. The presence of delta antigen in the blood may be associated with a depression of the levels of HBsAg, and even disappearance of HBsAg for a short period is seen in chronic HBV carriers. Anti-HDV tests are available and are reliable for diagnosing chronic HDV infection.

Serological diagnosis of acute co-infection is difficult. HBsAg and IgM anti-HBc will be present as they are in acute infection with HBV alone. However, anti-HDV antibodies are often delayed or may not be detected.

WATER-BORNE AND UNBORN

28. HEPATITIS E VIRUS

Hepatitis E virus is a newly identified agent causing epidemic hepatitis, especially in the developing countries. It has also been called enterically transmitted non-A, non-B hepatitis.

M. Degré

Hepatitis E is a viral infection resulting in a self-limited, enterically transmitted acute necroinflammatory disease of the liver. The disease occurs most frequently in epidemic outbreaks, predominantly in young adults. The clinical features are similar to those caused by hepatitis A, except for a high mortality rate that is observed among women in the third trimester of pregnancy.

TRANSMISSION/INCUBATION PERIOD/CLINICAL FEATURES

The virus is present in the stools before the outbreak and during the early phase of the disease. It is mostly transmitted by the faecal–oral route, most likely predominantly by way of contaminated drinking water. The incubation period is approximately 40 days (15–60 days).

SYMPTOMS AND SIGNS

Systemic:	Fever, Malaise, Anorexia
Local:	Jaundice, Abdominal Discomfort, Dark Urine, Pale Stools

Preicteric prodromal symptoms may be present for a few days and subclinical forms of the disease exist. Fulminant disease with high fatality rate in pregnant women has been reported in several outbreaks. Chronic liver disease has not been observed.

COMPLICATIONS

High mortality rate (10–20%) among infected pregnant women, especially those in their third trimester, has been observed in several major epidemics. High incidence of disseminated intravascular coagulation has been noted in association with the disease.

THERAPY AND PROPHYLAXIS

There is no specific therapy available. Hygienic control of spread is difficult because transmission is most prominent before the outbreak of disease, and often associated with poor sanitary conditions and shortage of clean water. It is not known whether immunoglobulin has any prophylactic effect.

LABORATORY DIAGNOSIS

Diagnosis is made by demonstrating the presence of IgM antibodies in the early phase of clinical disease, or demonstrating seroconversion. Identification of virus, viral antigens or viral nucleic acid in the faeces can be done, but it is not practical for clinical diagnosis.

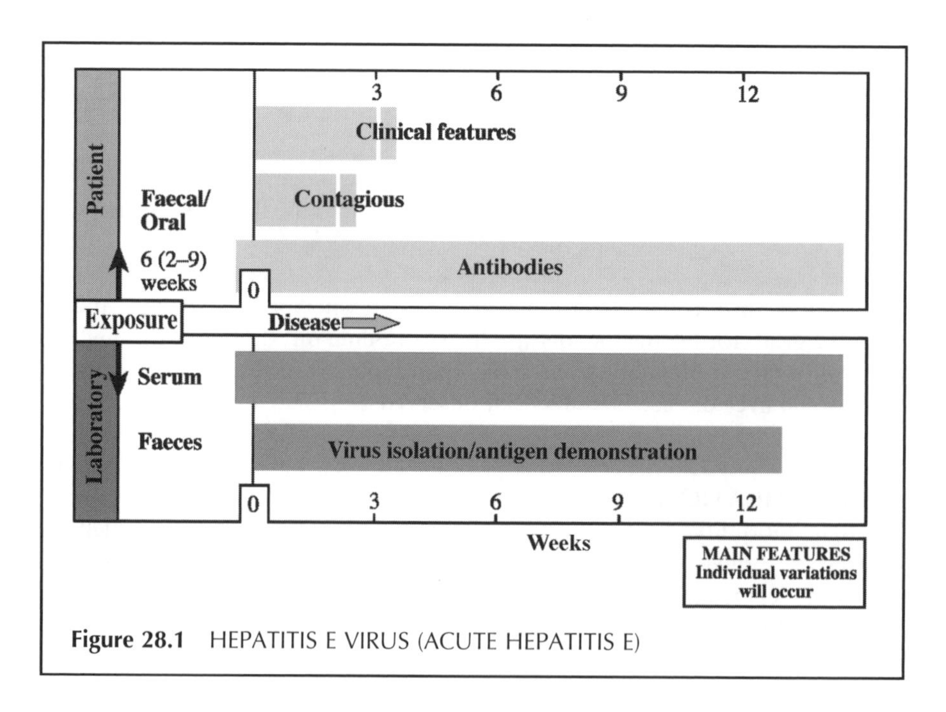

Figure 28.1 HEPATITIS E VIRUS (ACUTE HEPATITIS E)

CLINICAL FEATURES

SYMPTOMS AND SIGNS

The **incubation period** is approximately 6 (2–9) weeks. The disease may start with general symptoms, like fever, malaise, abdominal discomfort, but the regularity of such a prodromal phase has not yet been convincingly demonstrated. The main symptoms are those of a self-limiting acute jaundice, usually accompanied by general symptoms. In typical cases the urine becomes dark, the stools pale and there is an enlargement of the liver together with abnormal liver function tests. The development can be rather dramatic, especially among pregnant women. A 10–20% mortality rate in this group has been reported. The reason for this is not known. Chronic liver disease or persistent viraemia has not been observed.

Differential diagnosis. It is not possible to differentiate between hepatitis E and acute hepatitis caused by other infectious agents without the aid of laboratory tests. The history and epidemiological details are of great importance in reaching a presumptive diagnosis.

CLINICAL COURSE

Hepatitis E has symptoms of a self-limiting, acute icteric disease, similar to those caused by hepatitis A. Development of chronic hepatitis has not been observed. High mortality rate in pregnant women is often associated with massive hepatic necrosis.

COMPLICATIONS

Fulminant disease, especially in pregnant women, is described in several epidemics, with 10–20% lethality.

THE VIRUS

Hepatitis E virus particles can be recovered from the stools or from the bile of patients and infected monkeys. The HEV particles are spherical, non-enveloped with indentations and spikes on their surfaces with a diameter of 32–34 nm. They belong to the *Caliciviridae* family, although relation to the other members of the family is relatively distant. The genome is a single-stranded polyadenylated positive-sense RNA of approximately 7.5 kb. It contains three partially overlapping open reading frames (ORFs). These code for both structural and non-structural proteins, including several enzymes. They are most likely to be subjected to post-translational processing and modifications. Molecular sequencing of HEV clones revealed significant divergence between isolates from Asia and North America. Three genotypes can be differentiated. Although antigenic variations have been demonstrated, there is most likely a single serotype of HEV. The virus is relatively labile, both *in vivo* in the stools and *in vitro* during storage. HEV can be transmitted to chimpanzees and to several types of monkeys. *In vitro* culture has been reported.

EPIDEMIOLOGY

Hepatitis E occurs predominantly as an epidemic disease. Several major epidemics, involving several thousands of patients, have been described, mostly in Southeast Asia but also in central Asia, Africa and central America. Sporadic cases also occur in the same areas. Seroepidemiological studies in these areas indicated that up to 25% of the population in some countries may have been infected with HEV. A few cases have been reported from North America and Europe, generally in travellers. Seroepidemiological studies indicate that a low degree of circulation may occur in the industrialized countries, both in Europe and in North America. The source of the disease is in most instances faecally contaminated drinking water, and the epidemics are often observed after the rainy season. The highest attack rate is observed in young adults. A specially severe course of infection is recorded in pregnant women, with a high (10–20%) mortality. Secondary cases in families are infrequent.

THERAPY AND PROPHYLAXIS

There is no specific therapy available. Reinfection is rare with circulating antibodies indicating that antibodies to HEV protect against disease. However, to date no preventative effect of immunoglobulins has been demonstrated. General hygienic measures, including boiling of drinking water, can probably prevent infection.

LABORATORY DIAGNOSIS

IgM and IgA anti-HEV can be detected in serum up to 2 weeks before and 5–7 weeks after the onset of jaundice by EIA tests based on peptides produced in baculovirus or in other biological systems. IgG antibodies can be detected for at least 10 or more years after the acute disease. Specificity of a positive EIA test can be confirmed by Western blotting. Demonstration of the virus in the stools is possible by electron microscopy or PCR, but is not a practical diagnostic tool. The presence of HEV antigen in live biopsies can be shown by immunofluorescence.

SORTING OUT THE HEPATIC ALPHABET

29. EMERGING HEPATITIS VIRUSES

Other hepatitis viruses, non-A, non-B, non-C

G. L. Davis

Evidence that agents other than the commonly described hepatotrophic viruses can cause hepatitis comes from the observation of multiple distinct episodes of serologically uncharacterized hepatitis in some patients and the persistence of some cases of post-transfusion hepatitis despite elimination of hepatitis B and C from the donor pool. Additionally, most cases of non-A, non-B hepatitis are also due to HCV.

THE AGENTS

Several viral agents have been identified recently but these are unlikely to cause clinical liver disease. The "Hepatitis G" or "Hepatitis GB" is another flavivirus-like virus with 20–30% sequence homologous to HCV. Though probably transfusion transmitted, it is not felt to result in clinical disease and is no longer felt to be a hepatitis virus. The SEN and TT viruses are common in patients with hepatitis B and C, as well as the general population. Although solid evidence that they cause hepatitis is lacking, it does appear that they may be responsible for some cases.

TRANSMISSION/INCUBATION PERIOD/CLINICAL FEATURES

All 3 of the above-mentioned agents are common in transfusion recipients and patients with other forms of parenterally transmitted hepatitis, suggesting that the routes of transmission may be similar. It still remains unclear how prevalent these infections are and whether they account for an appreciable proportion of patients, if any, with cryptogenic hepatitis and cirrhosis.

THERAPY

Interestingly, retrospective testing of HCV co-infected subjects has shown that all 3 agents respond to interferon. Nonetheless, since liver disease as a result of these infections has not been proven, treatment should not be considered.

LABORATORY DIAGNOSIS

The lack of widely available serological markers for these infections has prevented identification of patients outside of research surveys. The studies reported to date have screened serum samples with either RT-PCR amplification of the viral RNA or an immunoperoxidase test to detect antibody.

FIFTH DISEASE?

30. PARVOVIRUS B19

Human parvovirus, human serum parvovirus. Erythema infectiosum. Slapped cheek disease. Fifth disease.

J. R. Pattison

Parvovirus B19 is a moderately contagious virus. The common result of infection is erythema infectiosum (EI). When the virus infects individuals with chronic haemolytic anaemia, aplastic crisis results.

TRANSMISSION/INCUBATION PERIOD/CLINICAL FEATURES

Transmission occurs by droplet inhalation. The interval between exposure and EI is 2.5–3 weeks. Patients are infectious for approximately 7 days, commencing 1 week after acquiring infection. Patients with EI are rarely infectious.

SYMPTOMS AND SIGNS

Systemic:	Prodromes: Fever, Chills
Local:	Rash, Arthropathy
Other:	*See* Complications

The rash persists for up to 1 week, but may recrudesce. Arthropathy and arthralgia may persist for weeks.

COMPLICATIONS

Aplastic crisis, fetal infection, chronic infection in the immunocompromised.

THERAPY AND PROPHYLAXIS

Intravenous immunoglobulin to control chronic infection in immunocompromised individuals. No vaccine.

LABORATORY DIAGNOSIS

In **erythema infectiosum** most patients have detectable IgM antibody at presentation. Aplastic crisis patients present towards the end of the period of viraemia. Virus is detectable in 60% of patients at presentation. In the remainder IgM is detectable. In **fetal infection**, damage may not be apparent until some weeks after maternal infection, thus detection of IgM in maternal blood is infrequent. Fetal infection may only be diagnosed by detection of virus in fetal blood or tissues.

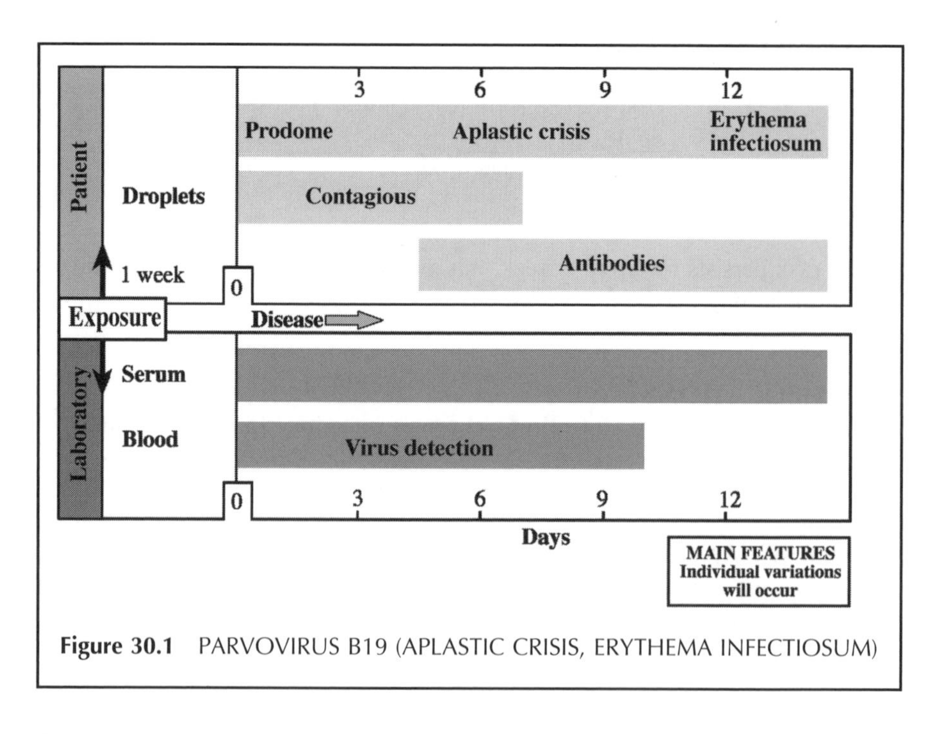

Figure 30.1 PARVOVIRUS B19 (APLASTIC CRISIS, ERYTHEMA INFECTIOSUM)

CLINICAL FEATURES

SYMPTOMS AND SIGNS

Transmission is commonly by droplet inhalation. Rarely, the transfusion of blood or blood products may lead to infection. At the peak of the viraemia, 1 week after exposure, there may be a mild flu-like illness with fever, chills and malaise, lasting 2–3 days. In many cases infection occurs without further symptoms. In **erythema infectiosum** the first sign is often marked erythema of the cheeks (slapped cheek appearance). A faint pink macular or maculopapular rash then develops on the trunk and limbs. As the rash fades a lacy or reticulate pattern may emerge. Recrudescence is common, especially after bathing or exposure to sunlight. In adult women symmetrical arthropathy is common (80% of cases), involving ankles, knees, wrists and fingers. This is normally resolved within 2–4 weeks. Low-grade fever may be present, and the rash may be pruritic. Leukopenia, reticulocytopenia and thrombocytopenia occur.

Differential diagnosis. Other exanthems, especially rubella, also scarlet fever, echovirus and coxsackievirus infections, and allergy. The diagnosis can only be made with certainty by laboratory tests.

CLINICAL COURSE

The rash persists for 4–5 days, although recrudescence may occur over 1 month. Arthropathy in adults resolves in the majority within 2–4 weeks, although in some arthralgia may persist for 2 months or so. Where arthritis occurs in children it is rarely symmetric, and may be severe for 1–2 months.

COMPLICATIONS

Aplastic crisis. The cell receptor for parvovirus B19 is the P antigen present on red blood cells, vascular endothelium and fetal myocytes. During all B19 infections, red blood cell precursors are infected in the bone marrow, causing a transient arrest of erythropoiesis for 7–10 days. In patients whose red cells have a shorter lifespan than normal, this arrest causes a rapid decrease in haemoglobin to very low, sometimes life-threatening, levels. Such aplastic crises occur in patients with hereditary haemolytic anaemias (e.g. sickle-cell anaemia). Aplastic crisis is treated by transfusion of packed red cells to raise haemoglobin levels.

Fetal infection. If infection occurs during pregnancy, the virus may cross the placenta. This occurs in just under one-third of cases of maternal B19 infection. In 90% of cases of maternal infection the pregnancy proceeds to term and to date no defects have been noted in such babies. However, about 10% end in spontaneous abortion in the second trimester (a 10-fold increase compared with controls). Occasionally infection during pregnancy manifests as hydrops fetalis.

Immunocompromised patients. Anaemia due to persistent infection may occur in patients who are immunocompromised due to acute lymphatic leukaemia, and HIV positivity.

THE VIRUS

Parvovirus B19 (Figure 30.2) is a single-stranded DNA virus of the genus *Parvovirus*, family *Parvoviridae*. The capsid is a naked icosahedron, 21 nm in diameter for which the full three-dimensional structure has been elucidated. It is composed of two structural proteins VP1 and VP2, the latter constituting approximately 80% of the total protein mass of the virus. The genome of B19 is a single-stranded DNA 5.5 kb long. It has a characteristic parvovirus structure with a linear coding region bounded at each end by terminal palindromic sequences that fold into hairpin duplexes. The virus packages plus and minus DNA strands into separate virions in approximately equal proportions, but the coding regions of B19 are confined to the plus strand. There are two open reading frames both driven by a single promotor at map position 6.

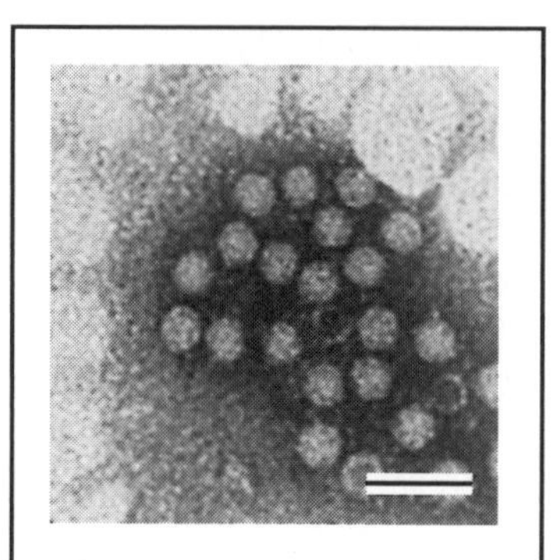

Figure 30.2 PARVOVIRUS B19. Bar, 50 nm (Electron micrograph courtesy of M. J. Anderson)

There appears to be a single stable antigenic type of B19 so that infection is followed by long-lasting immunity in an individual. However, genomic variants do occur and these have been classified into a number of groups. The groups do not correlate with the variety of clinical manifestations of B19, but individual viruses in the groups cluster in time and geographically.

The virus cannot easily be propagated in tissue culture, although replication occurs in primary cultures of human bone marrow. Virus replication requires host-cell function(s) found in late S phase of cell division.

EPIDEMIOLOGY

B19 virus is transmitted by droplet inhalation. Patients are infectious for about 1 week, coinciding with prodromal illness. Case-to-case intervals vary between 7 and 14 days. Infection is most common in primary school children aged 5–13 years, among whom outbreaks may occur. Evidence of past infection is present in 60–70% of adults. The virus is most common in the late winter and spring months. Epidemics occur every 3–5 years. The virus occurs throughout the world.

THERAPY AND PROPHYLAXIS

Due to difficulty in propagating the virus *in vitro*, there is no **vaccine**. During local epidemics, transfusion of packed erythrocytes or the administration of normal human **immunoglobulin** may afford some protection to individuals at risk of aplastic crisis. There is no indication for termination of pregnancy if maternal infection occurs, since no abnormalities have been noted in babies born to mothers whose pregnancy has been complicated by B19 infection. The symptoms of B19 infection may be treated as appropriate: erythrocyte transfusion for aplastic crisis, calomine lotion for pruritic rash, anti-inflammatory agents for arthritis. Intravenous immunoglobulin is effective in controlling chronic infection in the immunocompromised.

LABORATORY DIAGNOSIS

Erythema infectiosum. In the majority of cases specific IgM is present at the onset of the rash. This may be detected by antibody capture RIA or ELISA. Where the acute serum contains no detectable IgM, either a second sample taken 10 days later should be similarly examined, or the acute sample examined for virus (see below).

Aplastic crisis. Sera taken within 3 days of the onset of symptoms are likely to contain virus. Virus antigen may be detected by RIA or ELISA. Virus may also be detected by hybridization to cloned viral DNA labelled with biotin or ^{32}P. RIA, ELISA and DNA hybridization give positive results in 60% of acute sera from cases of aplastic crisis; this figure falls to 30% if CIE and/or electron microscopy are used.

Fetal infection. Diagnosis can only be made by examination of fetal material. It is most useful to test fetal blood for virus antigen or DNA, since the fetus may be too immature to synthesize IgM. DNA extracted from fetal tissue is tested by DNA hybridization for viral genome. Alternatively, formalin-fixed paraffin-embedded tissue may be examined for viral DNA by *in situ* hybridization.

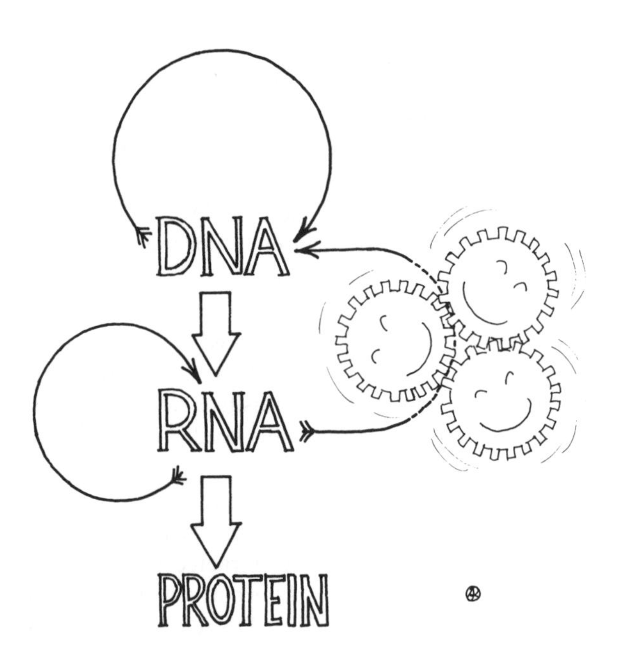

A STEP IN THE WRONG DIRECTION?

31. RETROVIRUSES

re = reverse; tr = transcriptase.

A. B. Dalen

Retroviruses are widely distributed throughout the vertebrates and perhaps the invertebrates. The main characteristics of this group of viruses is the presence in all infectious virions of an enzyme, reverse transcriptase, which catalyses the formation of a complementary DNA strand to an RNA template. A double-stranded DNA copy of the viral RNA genome (proviral DNA) may then be integrated and propagated in the host-cell genome. The chemical, physical and structural properties of the retroviruses from different species are essentially similar.

Retroviruses are enveloped RNA viruses, 80–100 nm in diameter. Most retroviruses carry surface glycoprotein spikes, anchored to the interior by a transmembrane protein or glycoprotein. The loosely arranged interior core contains a major and several minor proteins (the gag proteins). The viral genome consists of two identical strands of RNA to which reverse transcriptase is bound. The plasticity of the genome readily allows retroviruses to adapt to selective pressure. Point mutations occur with the same frequency as in other RNA viruses, but retroviral genomes recombine at an exceptional rate with other viral genes or host-cell sequences.

Endogenous retroviruses or retrovirus genomic material (retrogenes) are normal constituents of cells and are inherited as Mendelian elements. They may be expressed under certain circumstances as complete, infectious virus particles (exogenous retroviruses). Mostly, however, the gene sequence contains deletions and therefore remains unexpressed or gives rise to incomplete particles. Endogenous retroviruses are mostly animal viruses which may be 'rescued' when the cell is stressed in some way.

Exogenous retroviruses comprise three subfamilies:

- *Oncovirinae* (RNA tumour virus group). These are further subdivided on morphological criteria into: (a) intracellular A-particles, (b) B-type retroviruses (mouse mammary tumour virus), (c) C-type retroviruses (leukaemia viruses HTLV-1 and 2) and (d) D-type retroviruses (associated with primate neoplasias).
- *Lentivirinae* (slow virus group). Lentiviruses are non-oncogenic and cause chronic, inflammatory disorders in the host. HIV-1 and 2 are related to this group.
- *Spumavirinae* (foamy virus group). Spumavirus has been isolated from many species, including man, and has no known pathogenic potential.

MODE OF INFECTION AND DISEASE MECHANISMS

Retroviruses have generally low infectivity, and transmission under natural conditions usually requires intimate contact with blood, sperm, milk or sputum. Retrovirus infections are often but not invariably permanent, and the immune defence is in many systems inadequate to cope with the infection. Many cell types may be infected within the host. Some cells propagate virus without impairment of cellular function, while others may undergo malignant transformation or present other signs of cellular dysfunction.

The latent period before disease develops is usually very long in retroviral infections. Numerous organ systems may be afflicted, but the virus has a predilection for lymphocytes (leukaemias and lymphocyte destruction) and the central nervous system (chronic inflammatory disease). The mechanisms underlying cellular dysfunction caused by retroviral infections are poorly understood with the exception of acute transforming retroviruses.

Endogenous retroviruses have an aetiological role in certain neoplasias and inflammatory disorders in animals. A similar function has not been clearly demonstrated in the human counterpart.

A GLOBAL CHALLENGE

32. HUMAN IMMUNODEFICIENCY VIRUS (HIV)

AIDS = acquired immune deficiency syndrome. Fr. *SIDA.*

B. Åsjö

AIDS is the end-stage manifestation of human immunodeficiency virus (HIV) infection. Earlier phases of the infection are asymptomatic, persistent generalized lymphadenopathy (PGL), HIV-related symptoms and constitutional or neurological manifestations.

TRANSMISSION/INCUBATION PERIOD/CLINICAL FEATURES

Transmission is by sexual intercourse (especially homosexual) and by inoculation of infectious blood or blood products. The virus can be transmitted from mother to child, transplacentally, during birth or via breast feeding. Time from exposure to HIV until onset of acute clinical illness is 1–4 weeks. After primary infection there is a period ranging from a few months to more than 10 years with no or mild symptoms before the appearance of severe immunodeficiency. The patient is infectious before appearance of antibodies and remains so.

SYMPTOMS AND SIGNS

'Primary infection':	Influenza- or Mononucleosis-like, Erythematous Maculopapular Rash
HIV-related symptoms:	Canidida, Dermatological Symptoms, Varicella-Zoster
Constitutional symptoms:	Fever, Night Sweat, Fatigue
AIDS:	Opportunistic Infections, Kaposi's Sarcoma, Neuropathy, Dementia, Tuberculosis

The average survival time after the diagnosis of AIDS has been made is about 1 year if no antiviral therapy is given.

THERAPY AND PROPHYLAXIS

Various antiretroviral combination therapies are beneficial. There is no specific immunoglobulin or vaccine. Both clinical trials and clinical practice have convincingly demonstrated a significant reduction in mother-to-child HIV transmission if zidovudine is given as prophylaxis during pregnancy or nevirapine is given as a single-dose during delivery.

LABORATORY DIAGNOSIS

Specific antibodies appear 2–4 weeks to 3 months after infection and remain for life. Viraemia is detected at high levels before seroconversion. It is present at moderate levels during the clinically asymptomatic period. Low numbers of CD4+ T-cells in the blood are characteristic but not pathognomonic of HIV-associated immunodeficiency.

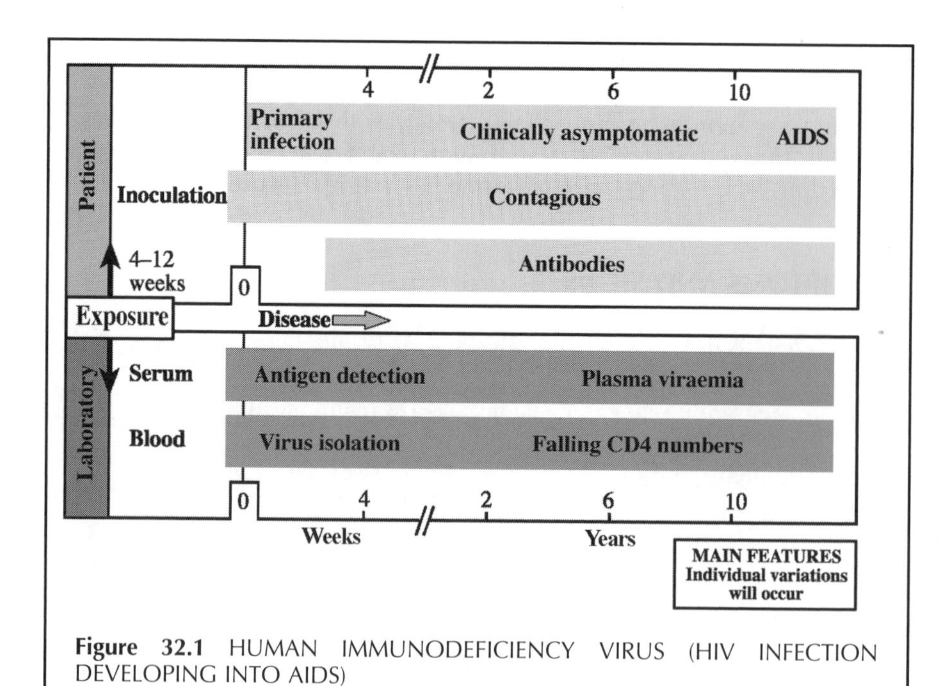

Figure 32.1 HUMAN IMMUNODEFICIENCY VIRUS (HIV INFECTION DEVELOPING INTO AIDS)

CLINICAL FEATURES

A clinical staging and classification system for HIV disease has been worked out by the Centers for Disease Control and Prevention, USA.

CATEGORY A includes:

Acute infection resembling mononucleosis or influenza with an erythematous maculopapular rash that may be seen in about 50% of cases 1–4 weeks after infection. Aseptic meningitis and encephalopathy, have been described in some cases. These acute manifestations are self-limited, but the HIV infection is persistent and chronic disease manifestations may develop later. The *differential diagnosis* includes primarily similar syndromes caused by EBV, CMV and *Toxoplasma gondii*.

Asymptomatic HIV infection, is defined as the time when the presence of HIV only can be demonstrated by laboratory analyses.

Persistent generalized lymphadenopathy (PGL), defined as swollen lymph nodes (at least 1 cm in diameter) in two or more non-contagious extra-inguinal sites, persisting for at least 3 months in the absence of any other illness or medication known to cause enlarged nodes. The lymphadenopathy may be associated with other manifestations of chronic HIV infection.

CATEGORY B consists of symptomatic conditions:

HIV-related clinical symptoms are indicative of a defect in cell-mediated immunity and often manifest as candidiasis (oral thrush), seborrhoeic dermatitis, hairy leukoplakia, yellow nails (fungus) and multidermatomal varicella-zoster.

Constitutional symptoms such as fever, diarrhoea, night sweat, fatigue and malaise can be seen in patients lacking criteria for AIDS definition. The term 'slim disease' is used in certain African countries for chronic HIV infections.

Neurological diseases (myelopathy and peripheral neuropathy) caused by HIV are seen in some patients lacking criteria for AIDS. In about 25% of patients a subacute encephalitis termed AIDS-dementia complex is manifested. Cerebral toxoplasmosis is here an important differential diagnosis.

CATEGORY C includes clinical conditions listed in the AIDS surveillance case definition:

AIDS is the term reserved for a person with at least one life-threatening opportunistic infection or Kaposi's sarcoma, with no identifiable reason for profound immunodeficiency and having a positive test for HIV infection. The average time to AIDS is 7–11 years if no therapy is given.

The most common opportunistic infections are pneumonia caused by *Pneumocystis carinii*, typical and atypical mycobacterial infections, often located extrapulmonary and disseminated, candida oesophagitis and severe HSV and varicella-zoster virus infections. AIDS patients may also present with severe wasting and the other symptoms and signs seen in HIV infections in general, including AIDS-dementia. AIDS inevitably leads to death in the course of less than a year in untreated cases.

THE VIRUS

HIV (Figure 32.2) is a complex RNA virus of the genus *Lentivirus* within the *Retroviridae* family. The virus is an approximately 100 nm icosahedral structure with 72 external spikes that are formed by the two major envelope glycoproteins gp120 and gp41. The lipid bilayer is also studded with a number of host-cell proteins during the budding process. HIV has a characteristic dense, cone-shaped nucleocapsid composed of the core protein p24. This nucleocapsid harbours two copies of the 9.8 kb single-stranded RNA genome which are associated with the viral enzymes reverse transcriptase (RT), RNase H, integrase and protease. In addition to structural genes HIV has genes whose products contribute to the complex regulation and replication of the virus. Of particular interest is the Nef (negative factor) protein. Deletions and mutations of this protein have been found in some HIV-infected individuals characterized as long-term non-progressors. Two major types of the AIDS virus, HIV-1 and HIV-2, have been identified. The major serological differences reside in the surface protein gp120. HIV-1 and HIV-2 are further separated into subtypes or 'clades' due to the marked variability in the V3 (variable region) of the gp120 protein. Infectivity is destroyed by most common disinfectants (hypochlorites, glutaraldehyde) and by heat.

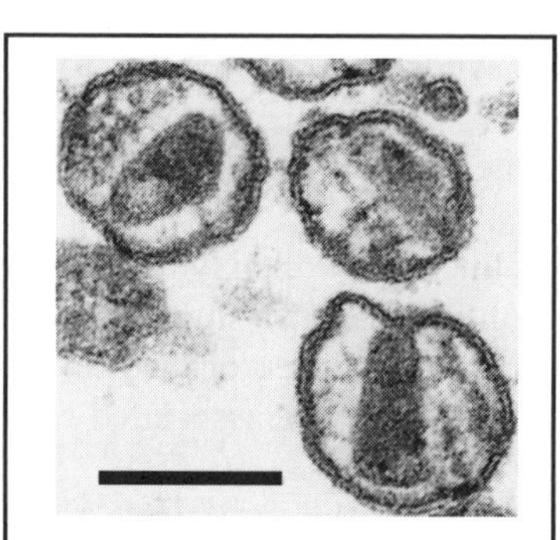

Figure 32.2 HUMAN IMMUNODEFICIENCY VIRUS. Bar, 100 nm (Electron micrograph courtesy of D. Hockley)

EPIDEMIOLOGY

AIDS was identified as a new disease entity in 1981 in the USA. The disease spread rapidly, first in the urban population on the east and west coast, later to all parts of the country and to other continents. Although the incidence of AIDS in the homosexual population has received great public attention and still accounts for the majority of AIDS cases in the USA and in Western Europe, worldwide heterosexual transmission is the leading route for the HIV pandemic;

in particular contacts with prostitutes, bisexual men and intravenous drug users. Sub-Saharan Africa, India, Thailand, the Russian Federation and China represent epidemic hot spots with rapidly increasing HIV prevalence. HIV is also transmitted from mother to child, *in utero*, intrapartum or perinatally via breast feeding. Transmission via whole blood or blood products has virtually ceased in industrialized countries after the introduction of blood screening. However, this is still a major concern in developing countries.

HIV-2 has been found in West Africa and in persons with sexual contact with West Africans. However, HIV-2 infections have been increasing rapidly in India. Interestingly, the disease process is slower than for HIV-1.

The total number of individuals living with HIV by the end of the year 2001 was reported by the World Health Organization to be 40 million. Of these, 25 million (70%) are living in sub-Saharan Africa. During the the year 2001 it was estimated there were 1600 new HIV infections per day worldwide, which corresponds to 10 infections per minute.

THERAPY AND PROPHYLAXIS

No curative **therapy** for HIV infection is available. Both clinical trials and clinical practice have convincingly demonstrated a significant reduction in mother-to-child HIV transmission if zidovudine is given as prophylaxis during pregnancy or nevirapine is given as a single-dose during delivery. Prophylaxis against AIDS-defining *P. carinii* pneumonia (PCP) has considerably improved the quality of life and prolonged the time before AIDS is diagnosed.

The introduction of the combination **antiviral therapy** regimens that contain at least one protease inhibitor with two reverse transcriptase inhibitors (see Chapter 4) has had a striking impact on AIDS lethality, with a dramatic decrease in the frequency of opportunistic infections. Though many patients experience improvements in quality of life with this therapy, there are several concerns with it. In particular, difficulties for many patients to adhere to the demanding drug dose schedules, side-effects, toxicities, development of multiresistant viruses and drug interactions. A key issue of intensive antiviral chemotherapy is to what extent the immune abnormalities induced by HIV can be reversed by efficient viral suppression and how long the beneficial effects last until eventual AIDS diagnosis. Guidelines for antiviral chemotherapy of HIV-infected adults, children and pregnant women have been published by the Department of Health and Human Services, USA and the British HIV Association.

No **vaccine** is available against HIV infection. A number of potential HIV vaccine candidates are under clinical trials. The obstacles are still numerous with respect to choice of vaccine strain, efficacy, target groups and behavioural changes among the vaccinees. At the present time the only effective prophylaxis is intensive, widespread and persistent information on the modes of transmission of HIV infection and risk-associated behaviour.

LABORATORY DIAGNOSIS

Tests to identify HIV infection can be divided into different categories: virus cultivation, antigen detection and viral genome amplification (PCR). Virus can be isolated from infected persons in most phases of the infection. Peripheral blood mononuclear cells (PBMC) can be co-cultivated with activated PBMC from HIV-negative donors in the presence of IL-2. A positive result is recognized by appearance of virus antigen (p 24) or reverse transcriptase activity in the culture medium.

Antibodies usually become detectable from 3 to 12 weeks after infection. As a rule, an infected person remains antibody-positive for life, but antibody titres often fall in patients with AIDS. The most widely applied tests are the indirect and the competitive ELISA, using mostly a mixture of viral antigens. It is recommended that confirmatory tests are carried out to exclude the possibility of false positive results. These are either variations of ELISA tests or Western blot analysis of antibody specificity.

The PCR technique represents a major advance in the diagnosis of HIV infection. This powerful technique can amplify target DNA present in minute amounts. It is therefore useful for early detection of HIV in infants born to infected mothers, since the presence of maternal IgG antibodies excludes serological testing during the first months after birth. Quantitative determination of plasma viraemia (virion RNA) by reverse transcription PCR (RT-PCR) has become a major tool to follow the progression of HIV infection in untreated patients and to monitor the effects of antiviral chemotherapy in patients.

A hallmark of chronic HIV infection is the depletion of CD4+ lymphocytes and loss of these cells is closely associated with acquisition of the characteristic opportunistic infections. The monitoring of CD4+ lymphocyte count is therefore an important determinant for clinical staging, initiation of antiviral therapy and PCP prophylaxis. However, the present knowledge that HIV is actively replicating in lymphatic organs throughout the infection raises the question of whether the peripheral blood, containing 2% of the total T-cell population, gives a representative picture of the pathogenic process. Plasma viraemia, together with CD4 counts, has therefore come to play a more important role in deciding when to initiate antiviral therapy.

ANOTHER ONE TO BE CHECKED?

33. HUMAN T-CELL LYMPHOTROPIC VIRUS TYPE I AND II

R. Whitley and G. Shaw

Human T-cell lymphotropic virus type I and II (HTLV-I and HTLV-II) are two known human transforming oncogenic retroviruses.

TRANSMISSION/INCUBATION PERIOD/CLINICAL FEATURES

HTLV-I has a low infectivity and is transmitted vertically from mother to child and horizontally by blood transfusions, intravenous drug abuse or sexually. Transmission by drug abuse of HTLV-I and HTLV-II is an increasing problem in many areas. HTLV-I exists in striking clusters in Japan, central and South America, Africa and the Caribbean Islands. The virus is rare outside the endemic areas. Clinical manifestations of early infection have not been delineated, although cutaneous manifestations and febrile disease have been reported. A few infected individuals (1 in 2000 to 1 in 100) develop adult T-cell leukaemia (ATL) after an average interval of 30 years.

SYMPTOMS AND SIGNS

HTLV-I:	Associated with Chronic Neurological Disease (HAM/TSP) and Adult T-Cell Leukaemia (ATL)
HTLV-II:	Not Yet Associated with Disease in Man

The typical aggressive form of ATL shows lymphadenopathy and frequent cutaneous and visceral involvement. An atypical, more indolent form of the disease has recently been reported.

Chronic progressive myelopathy (tropical spastic paraparesis, HTLV-associated myelopathy) has been identified as another clinical manifestation appearing with a low prevalence in HTLV-I seropositive individuals. Myelopathy and ATL very rarely coexist in the same patient. Variations in neurotropism of HTLV-I have been suggested. Host factors contributing to the development of myelopathy have not been identified.

COMPLICATIONS

See above.

THERAPY AND PROPHYLAXIS

No specific therapy available. Screening of blood and organ donors for antibody to HTLV-I has an impact on reducing seroconversions following administration of these products.

LABORATORY DIAGNOSIS

There is extensive cross-reactivity between HTLV-I and HTLV-II. Thus commercial ELISA tests cannot distinguish between these two agents. The use of PCR to peripheral blood cells is being performed in research laboratories.

CLINICAL FEATURES

SYMPTOMS AND SIGNS

No human disease has clearly been associated with HTLV-II, although it has been isolated from a case of hairy cell leukaemia. However, most hairy cell leukaemia cases are of B-cell phenotype. HTLV-I has been associated with chronic neurological disease, tropic spastic paraparesis (TSP) or HTLV-associated myelopathy (HAM), and adult T-cell leukaemia virus (ATL). Most HTLV-I infections are subclinical. ATL is a disease of early adulthood, mostly between the age of 20 and 30 years, appearing in approximately 1% of those infected, presumably through perinatal infection.

ATL is associated with acute infiltration of skin and visceral tissue with monoclonal proliferation of CD4 bearing T-lymphocytes. Clinical manifestations include skin lesions due to infiltrating leukaemic cells, interstitial pneumonia, hepatosplenomegaly and bone lesions. The entity of HAM or TSP is characterized by insidious onset and neurological findings of weakness, spasticity of the extremities, hyperreflexia, positive Babinski's sign, urinary and faecal incontinence, impotence and mild peripheral sensory loss. The disease is chronic and within 10 years 30% of patients are bedridden. Nevertheless, cognitive function remains normal.

CLINICAL COURSE

ATL is invariably fatal. HAM or TSP, on the other hand, is associated with progressive loss of neurological function.

COMPLICATIONS

Opportunistic infections are a common complication in individuals with ATL. Similarly, a paraneoplastics syndrome characterized by increased bone hypercalcaemia and lytic lesions is common.

THE VIRUS

HTLV-I and HTLV-II belong to the HTLV-BLV group within the *Retroviridae* family, being distinct from HIV-1 and HIV-2 (*Lentivirus*). Like other RNA retroviruses, the virion of HTLV-I consists of a viral envelope, an internal capsid core, the viral reverse transcriptase enzyme and the viral negative-sense RNA. The genomic structure of the integrated provirus DNA of HTLV-I and HTLV-II consists of two regions and the terminal repeats, a gag gene, which encodes for the internal core protein and the reverse transcriptase, followed by the envelope genes, encoding the glycoproteins gp21 and gp46. The HTLV has two unique genes (tax/rex; previously called X) coding for proteins activating transcription of the virus and probably various cellular proteins, including the receptor for IL-2. The development of ATL is a multistage process where HTLV-I is thought to initiate the first step, appearance of T-lymphocytes with an abnormal regulation of IL-2 receptors.

EPIDEMIOLOGY

Seroepidemiological studies of HTLV-I and HTLV-II show a geographical clustering of cases with the highest prevalence in Southern Japan, the Caribbean, Central Africa and the Southeastern USA. In Southern Japan, seroprevalence rates approach 35%. In the USA and Europe the seroprevalence of both viruses appears to be increasing in association with injected drug abuse and blood transfusion. In the USA HTLV-I and HTLV-II infect 7–49% of injected drug users. In the USA and Europe the viruses are most prevalent in intravenous drug users, whereas in the Caribbean sexual transmission appears to be the most common mode of spread; homosexually active men have a 5–10 times higher prevalence rate than the general population (15% vs. 2%). In addition, women who attend sexually transmitted diseases clinics and have other sexually transmitted diseases, such as syphilis, have higher rates of infection.

In addition to intravenous drug abuse and sexual contact, transmission is related to blood transfusion and breast feeding. A seroconversion rate of 63% has been observed in recipients of cellular components of contaminated units of blood. Recent data suggest that infection via breast milk is the primary route of HTLV-I transmission in Southern Japan. Persons who are seropositive for HTLV-I should be counselled regarding their inadequacy as blood and organ donors.

THERAPY AND PROPHYLAXIS

At present there is no therapy for HTLV-I and HTLV-II infections, nor are there currently any vaccines available. Introduction of routine screening of blood and organ donors for antibody is decreasing seroconversion rates.

LABORATORY DIAGNOSIS

Serological assays and molecular biological assays are needed to confirm the diagnosis of these oncoviruses. Commercial kits such as ELISA tests, Western blot strips or reagents for agglutination tests make screening for HTLV-I available. There is extensive cross-reactivity between HTLV-I and HTLV-II, making it difficult to distinguish between these two viruses. Some laboratories have developed viral peptides that will allow such distinction. Increasingly, PCR methods are used to detect HTLV-I and HTLV-II in peripheral blood and spinal fluid.

TICK BITES—A REAL PAIN

34. TICK-BORNE ENCEPHALITIS (TBE) VIRUS

Ger. *Frühsommer meningoenzephalitis*; Russian Spring–Summer (Eastern) encephalitis, Central European (Western) encephalitis.

T. Traavik

TRANSMISSION/INCUBATION PERIOD/CLINICAL FEATURES

Man is infected by the bite of the tick *Ixodes ricinus* (in Eastern Europe also *Ixodes persulcatus*), and consequently the disease is seen only in areas infested by these vectors. Their distribution is restricted by climatic conditions, vegetation and availability of host animals. TBE cases usually occur between April and November. The disease is seen in many European and Asian countries. Cases following consumption of infected, unpasteurized milk and cheese made from such milk are occasionally seen. Interhuman virus transmission has never been reported. The incubation time is 1–3 weeks, but may vary from 2 to 28 days. After milk-borne exposure the incubation period is shorter, 3–4 days.

SYMPTOMS AND SIGNS

Primary phase:	Moderate Fever, Headache, Myalgia
Secondary phase:	Flushing of Face and Neck, Conjunctival Infection, High Fever, Splitting Headache, Neck Rigidity, Vomiting, Pareses

Elevated temperature 2–20 days in the secondary phase. The two phases are usually separated by an afebrile and relatively asymptomatic period, lasting 2–10 days. The Eastern TBE subtype usually progresses without the intervening asymptomatic phase. Protracted convalescence is commonly seen. Reported case fatality rates are 0.5–2.0% for the Western and 5–20% for the Eastern subtype.

COMPLICATIONS

Chronic encephalitis with persistent CNS symptoms.

THERAPY AND PROPHYLAXIS

No specific therapy is available. Vaccines are available and have been used with high efficacy. In enzootic/endemic areas clothing may be used to avoid vector bites.

LABORATORY DIAGNOSIS

Serological diagnosis based on specific IgM detection is most practical and reliable. PCR tests for detection of TBE virus RNA in blood and CSF are being validated for clinical use.

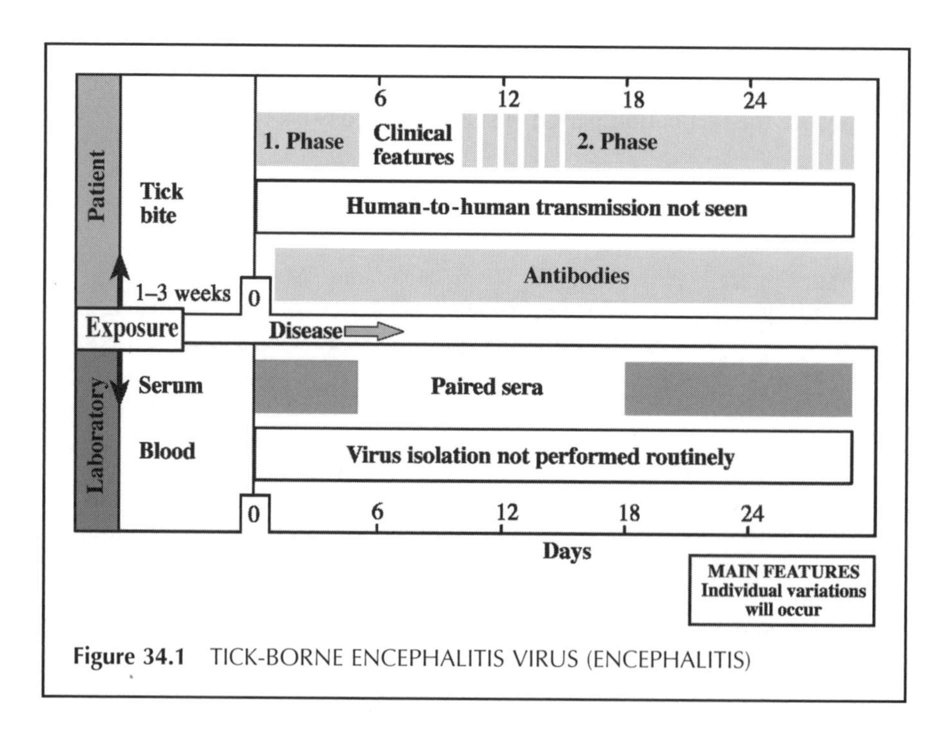

Figure 34.1 TICK-BORNE ENCEPHALITIS VIRUS (ENCEPHALITIS)

CLINICAL FEATURES

SYMPTOMS AND SIGNS

Man is infected following bite of the tick *I. ricinus* (and *I. persulcatus* for Eastern TBE). In highly infested areas such exposure will not always be recalled by the patient. The prodromal phase of TBE lasts 2–12 days and has no pathognomic symptoms. At this stage the disease will most commonly be diagnosed as a non-specific fever with headache and myalgia. If an EEG is performed, pathological findings may, however, be detected. The 'major disease' (secondary phase) may follow after a symptom-free period lasting 2–20 days. Some patients will present typical encephalitis symptoms, i.e. sleep disturbances, confusion, somnolence, vertigo or ataxia. Temporary pareses may be observed, not uncommonly in the shoulder region, but facial muscles, the pharynx, intestines and bladder may also be involved. In the cerebrospinal fluid (CSF) elevated protein concentration and a pleocytosis dominated by lymphocytes may be detected. In the peripheral blood a polymorphonuclear leukocytosis is often demonstrated. TBE should be considered in tick-infested areas at the time of the year when the vector is active.

The postencephalitic syndrome refers to the development, following acute TBE, of chronic symptoms and signs. This may occur in a significant proportion of cases. A Swedish study cited 36% of patients during a median of 47 months of follow-up. A range of non-specific symptoms including headache, reduced stress tolerance, impaired memory, balance and coordination disturbance, hearing loss, tremor and chronic fatigue have been described.

Differential diagnosis. Clues are given by the biphasic course, perhaps combined with pareses and polymorphonuclear leukocytosis. It should be noted that Lyme disease, caused by *Borrelia burgdorferii*, may be transmitted by the same vectors as TBE. The same is the case for *Ehrlichia* species. Both may cause disease that is similar to TBE, and co-infections between TBE and the other two agents have been described. As Lyme disease may be cured by antibiotic therapy, care must be taken to differentiate, and also to consider co-infections.

Early in the TBE disease the CSF white blood cell count may be as high as 1000 per ml, with a predominance of neutrophils. This may be confused with CSF findings of bacterial meningitis. A precise virological diagnosis can only be made by laboratory tests.

CLINICAL COURSE

The lethality may be considerable in Eastern TBE, but is low in Western or Central European TBE. The temperature is elevated for 2–20 days in the secondary stage. During convalescence many patients complain of headache following mental or physical stress, sleep disturbances, and problems with concentration and memory.

COMPLICATIONS

TBE virus infection has been implicated in chronic encephalitis with persistent CNS symptoms.

THE VIRUS

TBE virus is classified within the genus *Flavivirus* of the family *Flaviviridae*. The family also comprises *Pestivirus* and hepatitis C viruses, and embraces small (40–50 nm) RNA viruses. The virus particle consists of a cubic capsid surrounded by a lipoprotein membrane (envelope). The RNA genome is single-stranded, unfragmented and has a positive polarity (i.e. mRNA sense, sized approximately 11 kb). The virion contains only three polypeptides. The capsid is formed by protein C, while the envelope contains two glycoproteins, named M and E. Eastern and Western subtypes can be distinguished by differences in the proteins, and strains with different virulence have been recognized within each subtype. Protein E is probably the most immunogenic protein, inducing neutralizing and protective antibodies. Some non-structural polypeptides have enzymatic functions necessary for productive infection. TBE viruses are heat-labile and inactivated by pasteurization. They survive in milk for prolonged periods, even during passage through the acidic environment of the stomach.

EPIDEMIOLOGY

TBE is endemic to Northern, Central and Eastern Europe, Russia and the Far East. Cases have been reported from Austria, Byelorussia, Bulgaria, China, the Czech Republic, Denmark, Estonia, Finland, France, Germany, Greece, Hungary, Italy, Japan, Kazakhstan, Latvia, Lithuania, Norway, Poland, Russia, Romania, Slovenia, Sweden and Switzerland. With increasing inter-continental travel to endemic areas, one would assume that the risk of tourists attracting TBE will increase, and a rise in incidence has indeed been observed recently.

The major host animals for ticks are small rodents and passerine birds, but larger animals may also be involved. Seroprevalence in large mammals is an indirect measure of TBE virus transmission intensity. The natural host animals do not seem to be clinically affected by TBE virus infections. TBE virus is able to pass transovarially from one infected female tick to her offspring. Consequently, the tick may be regarded not only as a vector, but also as a reservoir for TBE virus. The activity of ticks is temperature-dependent, and this is reflected in the seasonally recorded cases of TBE in different areas. The prevalence of ticks infected with TBE virus in endemic areas in Europe usually varies from 0.5% to 5%, while in some regions of Russia, a prevalence of 40% has been reported. Cases of human-to-human virus transmission have not been reported. However, infections have been observed upon ingestion of unpasteurized milk from viraemic animals. In some European countries TBE represents a medical problem. In others (for instance Norway) a high prevalence of seropositive individuals has been recorded, but most infections seem to run an abortive or subclinical course. No obvious reasons for this discrepancy have been found so far, but the occurrence of TBE virus strains with different virulence may be a reasonable explanation.

The highest incidences of TBE have been registered in Latvia, the Urals and the Western Siberian regions of Russia. The attack rates in these areas may reach 199 cases per 100,000 inhabitants per year. The incidence of subclinical TBE cases is difficult to establish, but data suggest rates of 70–98%.

The increase in TBE cases seen in the last century may be related to global warming. It is hypothesized that temperature increase leads to larger rodent populations and higher tick activity.

THERAPY AND PROPHYLAXIS

Treatment is supportive, since no **specific chemotherapy** or **immunotherapy** is available. Clinical observations suggest that bed rest for up to 2 weeks may improve the outcome. Studies evaluating symptomatic treatments have not been performed. Paracetamol and aspirin are often given in mild cases, and may give symptomatic relief. Patients with signs and symptoms of meningoencephalitis should be closely observed, because coma or neuromuscular paralysis leading to respiratory failure may develop rapidly, i.e. within 1 hour.

Active **vaccination** appears to be the most effective means to prevent TBE. The most widely used vaccine in Europe is FSME-Immune (Immuno AG, Vienna, Austria), which consists of whole purified, formaldehyde-inactivated virus. Following mass vaccination in Austria, a dramatic decline in the incidence of TBEV infection was demonstrated. An estimated vaccine efficacy of more than 95% has been reported. Naked DNA and also RNA vaccines are being developed at the moment. Otherwise, the only practical prophylactic measure is protective clothing to avoid tick bites, and immediate removal of ticks from the body.

LABORATORY DIAGNOSIS

During the prodromal stage TBE virus may be isolated from the blood, either by mouse inoculation or in cell culture, but serology is the best method of diagnosis. Patients with other flavivirus infections produce IgG antibodies that cross-react with TBEV. Specific diagnosis depends on the detection of anti-TBEV IgM antibodies, in serum or CSF, by ELISA. The tests are highly sensitive and specific. The timing is essential, especially when disease presents without an initial febrile phase. In suspected early cases, tests should be repeated 1 week after onset of fever. IgM antibodies in serum may be detectable for up to a year. Local synthesis of intrathecal antibodies usually correlates with serum levels.

RT-PCR detection of TBEV RNA has been successfully applied for some purposes, including viraemia in animal hosts. Routine clinical use still needs to be validated. RT-PCR based detection of TBEV sequences in brain tissue has been reported.

CALIFORNIA ENCEPHALITIS (CE) VIRUS

Viruses belonging to the antigenic group CE, genus *Bunyavirus* within the family *Bunyaviridae* (Figure 34.2). The CE viruses, as other bunyaviruses, are medium-sized (90 nm diameter), enveloped RNA viruses with spherical shape. The virions range from 80 to 120 nm in size. The envelope contains glycoprotein spikes and surrounds a core consisting of the RNA genome and its associated proteins. The RNA genome is mostly of negative polarity and divided into three segments: large (L), medium (M), and small (S). The size of the segments is approximately 7, 4.5 and 1 kb, respectively, and they are coding for defined structural and non-structural polypeptides. There may be overlaps in vector species and host animals between CE virus strains. It has therefore been hypothesized that new CE viruses might arise by reassortment of RNA fragments between different parental virus strains.

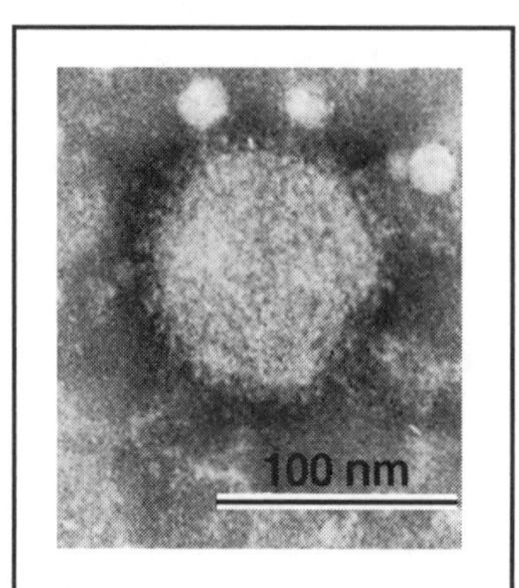

Figure 34.2 BUNYAVIRUS. Bar, 100 nm (Electron micrograph courtesy of E. Kjeldsberg)

The CE group includes about 15 virus strains distinct enough to deserve individual names. They are transmitted by mosquitoes, most commonly by *Aedes* or *Culiseta* spp., and each virus has a limited geographic distribution and a very narrow range of mosquito and mammalian hosts. The CE viruses cause human infections on five continents.

In the USA two different CE viruses (La Crosse and Jamestown Canyon) are of medical importance. Infections may result in relatively severe encephalitis. La Crosse infections cause 8–30% of all annual encephalitis cases in the USA. Paediatricians have been concerned about La Crosse virus infections due to reports of sequelae, i.e. mental retardation. In Central and Southern Europe Tahyna virus may cause influenza-like disease and less severe meningoencephalitis. In Finland, Norway and Sweden various CE virus strains, among them Inkoo virus, have been isolated from mosquitoes. High antibody prevalences have been demonstrated among humans, domestic animals and wildlife. Several cases of clinically silent seroconversions in humans have been documented. Inkoo virus has not been associated with human disease in Western Europe, but Russian studies have indicated that it can cause encephalitis. Specific sampling of potential vectors for virus isolation, detailed characterization of virus strains, and the use of fully characterized strains for serological diagnosis will help to elucidate the potential of CE group viruses as human pathogens in Europe and elsewhere.

OUTDOOR PURSUITS

35. HANTAVIRUSES—HFRS and HPS

Hantaan river runs along the 38th parallel in Korea.

D. Wiger

Hantaviruses (genus *Hantavirus*, family *Bunyaviridae*) are the causative agents of a disease complex named haemorrhagic fever with renal syndrome (HFRS) that occurs primarily in Eastern Asia and the Balkans. The milder form of the disease, nephropathia epidemica (NE), is seen in the northwestern parts of Eurasia, especially in Scandinavia. The hantavirus pulmonary syndrome (HPS) was first described in the southwestern USA but more recently has also been reported in other parts of the Americas.

TRANSMISSION/INCUBATION PERIOD/CLINICAL FEATURES

The infection takes place after inhalation of aerosols containing excretions/secretions from small rodents infected with the virus. The most important animal reservoir in Europe and the western parts of the former USSR is the bank vole (*Clethrionymus glareolus*). In the eastern parts of Eurasia the main reservoirs are field mice (*Apodemus* spp.) and rats (*Rattus norvegicus*, *R. rattus*). In the USA the deer mouse (*Peromyscus maniculatus*) has been associated with disease in humans. The incubation time is estimated to be 1–6 weeks. Human-to-human transmission had not been unequivocally demonstrated until interhuman spread of HPS in South America was reported.

SYMPTOMS AND SIGNS

HFRS (NE)

Systemic:	Acute High Fever, Myalgia, Headache, Thrombocytopenia, Shock*, Haemorrhagic* Manifestations
Renal:	Tenderness Over the Kidneys, Oliguria (Later Polyuria), Haematuria, Proteinuria
Others:	Abdominal Pains, Nausea, Vomiting, Transitory Myopia, Mental Confusion*

*More common in HFRS than NE

HPS	
Systemic:	Acute High Fever, Myalgia, Headache
Respiratory:	Cough, Acute Respiratory Distress
Others:	Nausea, Vomiting, Gastrointestinal Distress

Both HFRS and HPS start with influenza-like symptoms (acute high fever and myalgia). After 2–5 days patients develop haemorrhagic manifestations and/or renal symptoms. In milder cases of HFRS there may be few overt renal symptoms but signs of liver involvement may be present. Pulmonary symptoms dominate in HPS and the disease can progress very rapidly to fatal respiratory failure.

COMPLICATIONS

HFRS: Kidney failure is often observed with transitory reduced glomerular filtration rate that may persist for months after the acute phase of the disease.

THERAPY AND PROPHYLAXIS

The treatment in HFRS (NE) is symptomatic. Renal dialysis and support of homeostasis (fluid and electrolyte balance) are important. In HPS, respiratory and circulatory support, in addition to haemodialysis, may be necessary. The antiviral drug ribavirin may be effective in reducing mortality in severe cases of HFRS, but the effectiveness in treatment of HPS has not been established.

LABORATORY DIAGNOSIS

Attempts to isolate the virus from clinical specimens are rarely successful. For HFRS and NE serological diagnosis is based on the detection of IgM antibodies in acute phase serum or IgG titre rise in paired sera using indirect immunofluorescence. The disease progression in HPS is often so rapid that a virus-specific antibody response may not be detectable in the early acute phase of the disease. PCR tests for hantavirus RNA have been used successfully to confirm the presence of virus in tissue samples.

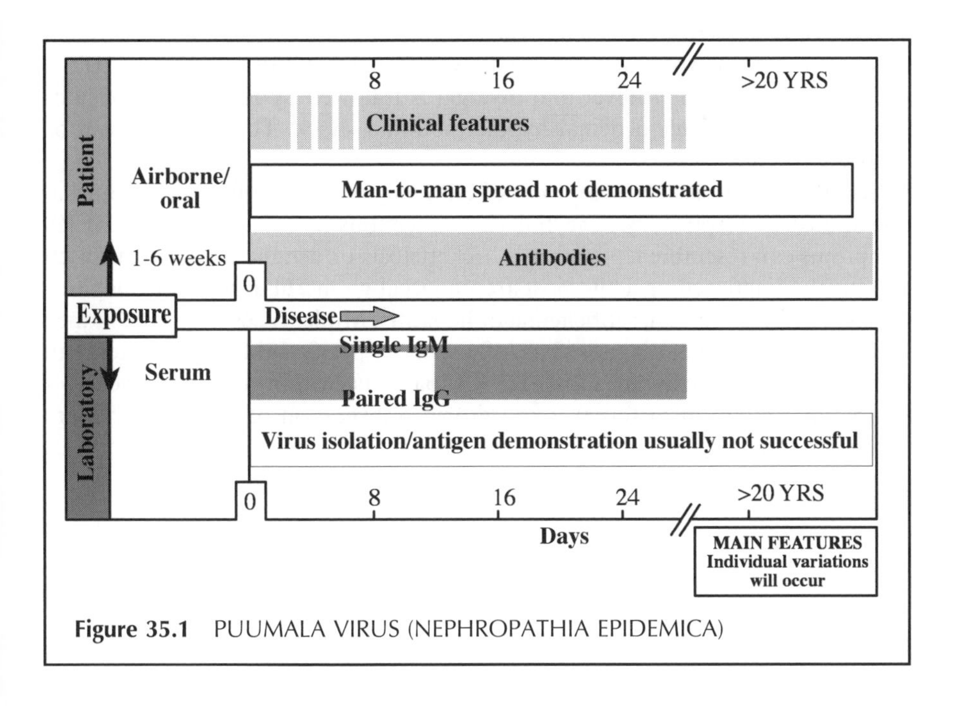

Figure 35.1 PUUMALA VIRUS (NEPHROPATHIA EPIDEMICA)

CLINICAL FEATURES

SYMPTOMS AND SIGNS

In HFRS acute fever over a period of 3–6 days is followed by a hypotensive phase which can develop into classical shock symptoms. An oliguric phase, when extensive haemorrhaging may occur, is then followed by a diuretic phase. Depending on the supportive therapy, mortality ranges from 5 to 20%. In the milder forms of the disease such as NE acute high fever, often with headache and malaise, is followed by abdominal/lumbar pain and renal involvement. Elevated levels of liver enzymes and increased SR and CRP are often observed. There is less tendency for haemorrhages, but in the more serious cases renal failure can develop that may require dialysis. The mortality is $<1\%$. In HPS a prodrome of acute high fever and myalgia is followed by rapid onset of non-cardiogenic pulmonary oedema, hypotension and shock. The mortality can be $>50\%$.

Differential diagnosis. The symptoms of HPS may resemble influenza, legionella, pneumonia or respiratory syncytial virus infections in young children. HFRS symptoms can resemble leptospirosis, rickettsiosis or dengue fever. NE is often tentatively diagnosed as acute poststreptococcal glomerular nephritis, nephrolithiasis, acute abdomen, septicaemia or hepatitis. Milder cases with little kidney involvement are often diagnosed as influenza. A specific IgM antibody test can confirm the diagnosis in most HFRS patients within the first week of symptoms (initiation of fever). Laboratory confirmation of HPS may require detection of hantavirus antigen or viral genomic fragments in tissue specimens.

THE VIRUS

Hantaviruses belong to the genus *Hantavirus* within the *Bunyaviridae* family. The medium-sized single-stranded negative-sense RNA genome consists of three unique segments: large (L) that codes for virus polymerase, medium (m) that codes for glycoproteins G1 and G2 and small (S) that codes for the nucleocapsid. There are at least three groups of hantavirus that have been connected with disease. Each of these virus groups is associated with specific rodent hosts: Hantaan-like virus with *Apodemus* spp., Seoul-like virus with *Rattus* spp. and Puumala-like virus with *Clethrionymus glareolus*. The virus implicated in most cases of HPS in the USA has been named Sin Nombre and is associated with *Peromyscus maniculatus*. However, in other areas of the Americas, different Sigmodontinae rodent reservoirs are important. The Sin Nombre-like viruses seems to be more related to Prospect Hill virus, which was isolated from *Microtus pennsylvanicus* in the USA. However, the Prospect Hill virus has not been associated with any human disease.

EPIDEMIOLOGY

In the northern hemisphere many species of small rodents (voles and field mice) experience dramatic population fluctuations with peaks every 3–4 years. The incidence of human infections increases during the years when rodents are most numerous. The seasonal variation is related to climatic conditions, the ecology of the animal reservoir and the activity of the human host. Most cases of HFRS are seen in the autumn and early winter when rodents invade human dwellings, or because of occupational and/or recreational activity when people come into contact with the natural habitat of the rodent. In the Balkans and other parts of Southern Europe, HFRS may also occur in the spring and summer months. There is a higher prevalence of males among patients with hantavirus infections. This may be related to the increased chance of exposure to the virus in traditionally male-dominated occupations or recreational activities.

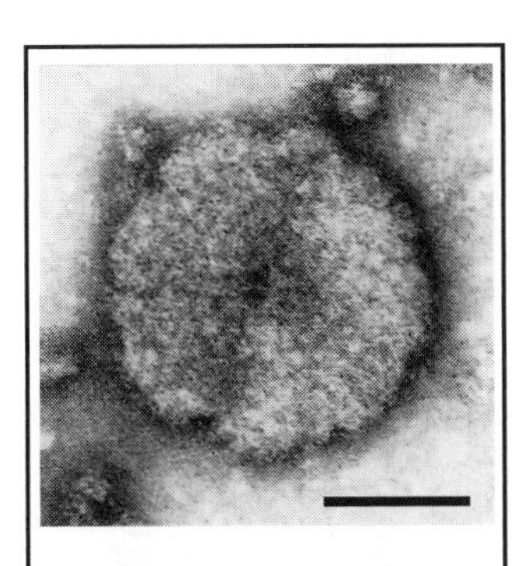

Figure 35.2 PUUMALA VIRUS. Bar, 100 nm (Electron micrograph courtesy of E. Kjeldsberg)

DIGGING THE PANAMA CANAL: MAN AGAINST YELLOW FEVER

36. HAEMORRHAGIC FEVER VIRUSES

G. Haukenes

Haemorrhagic fever viruses are found within several families. They have natural transmission cycles among animals and are therefore geographically restricted to certain areas.

FLAVIVIRUSES (Lat. *flavus* = yellow)

Flaviviruses are enveloped positive-sense single-stranded RNA viruses, about 50 nm in diameter. **Yellow fever** and **dengue** haemorrhagic fever occur in the tropics and adjacent subtropics and are transmitted by mosquitoes (*Aedes* and *Haemagogus*). The host reservoirs of yellow fever virus are humans (urban form) and monkeys (jungle form). **Kyasanur Forest** (India) disease and **Omsk** (Western Siberia, Russia) haemorrhagic fever are transmitted by ticks, the reservoirs being different vertebrate animal species.

Common clinical features are fever, headache, haemorrhages in the skin and mucous membranes. Case fatality rate is particularly high in secondary attacks of dengue, apparently due to immunological mechanisms. Deafness, hair loss and psychomotor disturbances are frequently recorded sequelae in Omsk haemorrhagic fever.

There is one distinct type of the yellow fever virus and four types of dengue fever virus. The Kyasanur disease and Omsk fever viruses belong to the TBE complex (see Chapter 34).

Interferon treatment seems to reduce lethality in dengue. The live (17D) yellow fever **vaccine** is safe and effective. Inactivated vaccines seem to provide some protection against Omsk haemorrhagic fever and Kyasanur disease. Important preventive measures are insect control and protection against bites.

Virus may be isolated from the blood in the acute phase of the disease. Serological diagnosis is performed by CF and HI and by ELISA for demonstration of specific IgG and IgM, with due attention to cross-reactions between members of the flavivirus family.

TOGAVIRUSES (Lat. *toga* = gown, cloak)

Togaviruses are somewhat larger than flaviviruses (70 nm in diameter), but have the same gross morphological structure (for more virus details, see Chapter 15). Haemorrhagic fever is caused by Chickungunya (East Africa) virus belonging to the *Alphavirus* genus. The disease, which is milder than dengue haemorrhagic fever, is seen in Asia and Africa and is transmitted from man to man by *Aedes aegypti* and other mosquitoes.

Table 36.1 HAEMORRHAGIC FEVER VIRUSES: MODE OF TRANSMISSION AND CURRENTLY AVAILABLE PROPHYLAXIS AND THERAPY

Disease	Transmission	Immunoglobulin/vaccine	Antiviral
Yellow f.*	Mosquito bite	Live vaccine	Interferon
Dengue h.f.	Mosquito bite		Interferon
Omsk h.f.	Tick bite	Inactivated vaccine	
Kyasanur Forest disease	Tick bite	Inactivated vaccine	
Chickungunya f.	Mosquito bite		
Rift Valley f.	Mosquito bite	Inactivated vaccine	
Crimean–Congo h.f.	Tick bite		
Lassa f. (Argentinian/ Bolivian h.f.)	Contact: rodents man	Immunoglobulin	Interferon Ribavirin
Marburg/ Ebola disease	Contact: rodents monkeys man	Convalescent plasma	Interferon Ribavirin

*(h.)f. = (haemorrhagic) fever.
Haemorrhagic fevers with renal syndrome are discussed in Chapter 35.

BUNYAVIRUSES (Bunyaamvera in Uganda)

Bunyaviruses are large, enveloped negative-sense single-stranded RNA viruses with a segmented genome. Haemorrhagic fever with renal syndrome (HFRS) is discussed in Chapter 35.

Rift Valley (East Africa) fever virus (genus *Phlebovirus*) is transmitted by mosquitoes (*Culex* and *Aedes*) from cattle to man, and occurs in Africa. Serious symptoms including encephalitis, blindness and haemorrhages are seen in some cases.

Crimean–Congo haemorrhagic fever virus (genus *Nairovirus*) is seen in Africa, Asia and Eastern Europe and is transmitted to man by ticks feeding on various wild and domestic animals. Man-to-man transmission also occurs.

An inactivated **vaccine** is available for Rift Valley fever. Otherwise prophylaxis consists of mosquito and tick control and protection against bites.

ARENAVIRUSES (Lat. *arenosus* = sandy)

Arenaviruses are pleomorphic enveloped single-stranded negative-sense RNA viruses with ribosome-like ('sandy') particles in the envelope. The natural host reservoir is in rodents which have acute or persistent infections.

Lassa fever is a severe haemorrhagic fever seen in rural West Africa. Similar fevers are **Argentinian** (Junin virus) and **Bolivian** (Machupo virus) haemorrhagic fevers. Transmission occurs on contact with secretions from infected rodents. In addition, virus may be transmitted by contact with patients.

Handling of viruses and clinical material from patients with Lassa fever should be restricted to special laboratories (containment category 4) equipped for work with highly infectious agents. Some therapeutic effect is obtained by the use of specific immunoglobulins, interferons and ribavirin (see Chapter 4). Rodent control is an important preventive measure in endemic areas.

FILOVIRUSES (Lat. *filo* = thread-like)

These viruses are enveloped negative-sense single-stranded RNA viruses forming filamentous elements of varying lengths up to 1400 nm. They cause the severe haemorrhagic fevers termed **Marburg** (from an outbreak associated with monkeys imported to Europe) and **Ebola** (a river in the Democratic Republic of Congo, formerly Zaire) diseases. The transmission cycles of these viruses are elusive, but person-to-person transmission seems to be important in human outbreaks. Serological surveys from Central Africa indicate that filoviruses commonly cause subclinical infections in these areas. The natural host is unknown.

Work with filoviruses causing haemorrhagic fever is restricted to containment category 4 laboratories. Treatment with interferon, convalescent plasma and ribavirin (see Chapter 4) seems promising. There is no vaccine against filoviruses.

ENRAGING THE WORLD

37. RABIES VIRUS

Gr. *hydrophobia*; Lat. *rabies* = madness; Ger. *Tollwut*; Fr. *rage*.

B. Bjorvatn and G. Haukenes

Rabies is a viral zoonosis. The virus reservoir is found among wild mammals, particularly foxes. Rabies virus causes an encephalomyelitis leading to degeneration of neurons in the central nervous system. The word 'rabies' reflects the frightening change of behaviour characteristic of the disease.

TRANSMISSION/INCUBATION PERIOD/CLINICAL FEATURES

Humans usually become infected through dog bites. Transmission of the infection is most efficient at the onset of the rabid period and for the subsequent 7–10 days. Human-to-human transmission is extremely rare. The incubation period may vary from 10 days to more than 1 year, but is in most cases 2–10 weeks.

SYMPTOMS AND SIGNS

'Furious' (rabid) rabies:	Hyperexcitability, Aggression, Convulsions, Hydrophobia
Paralytic (dumb) rabies:	Gradually Spreading Pareses

Progressive neurological symptoms and paralysis lead to death after 2–10 days.

THERAPY AND PROPHYLAXIS

After onset of clinical symptoms only symptomatic treatment is given. Postexposure vaccination, in serious cases combined with specific hyperimmunoglobulin, is highly effective. Pre-exposure vaccination is restricted to special risk groups.

LABORATORY DIAGNOSIS

Serological tests are of no value at the time of clinical symptoms. The diagnosis is verified by biopsies from skin, cornea impressions or post-mortem examination of brain tissue.

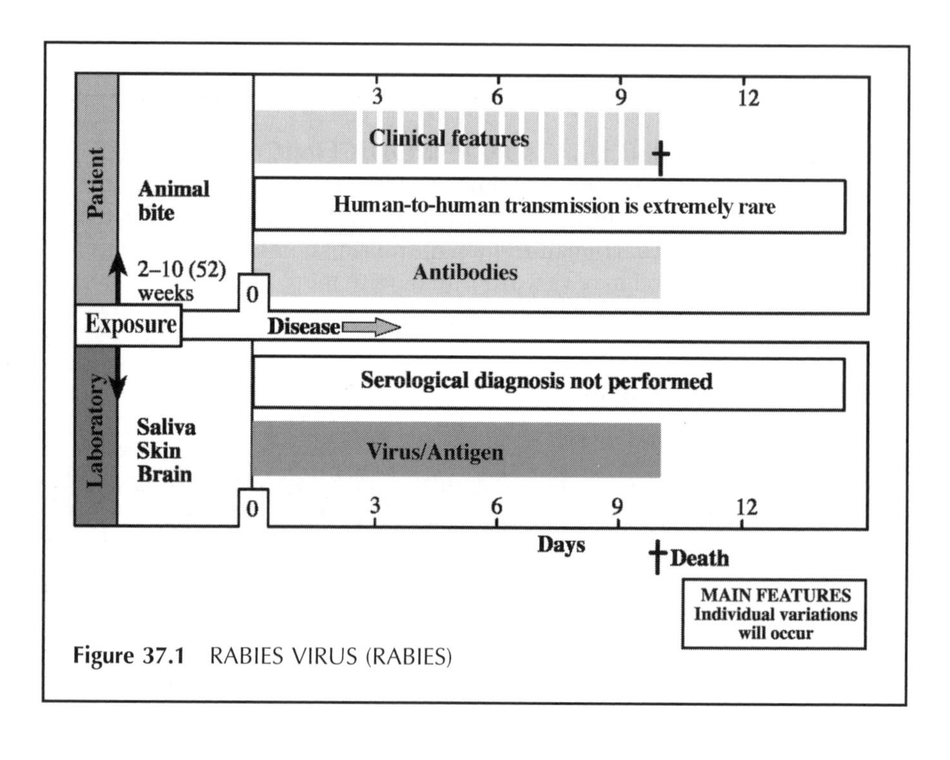

Figure 37.1 RABIES VIRUS (RABIES)

CLINICAL FEATURES

SYMPTOMS AND SIGNS

The incubation period is highly variable and may range from 10 days to more than 1 year, in most cases 2–10 weeks. The risk of becoming infected depends on the nature and location of the bite. Severe bites involving the head or neck are considered to carry the highest risk, and in such cases clinical disease may develop after a relatively short incubation period. In addition to site and severity of the bite, the virus concentration in the saliva of the animal is important. A few days before clinically manifest disease the patient may experience non-specific symptoms such as malaise, restlessness, insomnia and depression. The most common manifestation of rabies is the so-called **'furious' (rabid) rabies**, where the clinical picture is dominated by periods of intense hyperexcitability, hyperactivity and aggression, commonly combined with hyperventilation, hypersalivation and convulsions. Between such attacks the patient is usually anxious and exhausted, but well orientated. Hydrophobia as a result of painful spasms of the laryngeal muscles upon swallowing is commonly seen. The rabid phase is gradually replaced by spreading and increasing pareses, and finally coma ensues. The **paralytic (dumb)** form of rabies usually starts with prodromes as described above, followed by rapidly spreading paralysis. Combined forms are also seen.

Differential diagnosis. In the prodromal stage the diagnosis of rabies is usually not suspected unless there is a history of exposure. Other viral and post-infectious encephalitides have to be considered. Patients with rabies phobia may present with hydrophobia (rabies hysteria, pseudorabies), but they show no real hyperexcitability.

CLINICAL COURSE

The disease usually progresses to the terminal stage in the course of 2–10 days. Patients with paralytic (dumb) rabies often survive longer than rabid patients. The immediate cause of death is mostly a severe convulsion or respiratory muscle paralysis.

COMPLICATIONS

Various complications are seen, especially if the course is protracted. Most common are cardiovascular and respiratory disturbances.

THE VIRUS

Rabies virus (Figure 37.2) belongs to the genus *Lyssavirus* (Gr. *lyssa* = rage, rabies) within the *Rhabdoviridae* family (Gr. *rhabdos* = rod). The virus particle

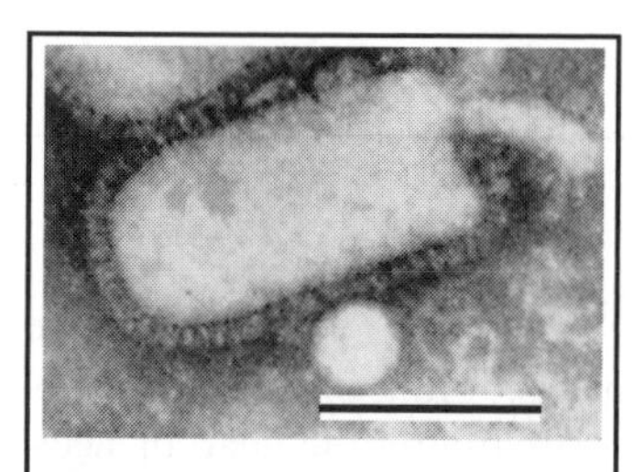

Figure 37.2 VESICULAR STOMATITIS VIRUS, A MEMBER OF THE RHABDOVIRUS FAMILY AND MORPHOLOGICALLY SIMILAR TO RABIES VIRUS. Bar, 100 nm (Electron micrograph courtesy of E. Kjeldsberg)

is cylindrical and bullet-shaped with dimensions of about 75×180 nm. The lipoprotein envelope has spikes of viral glycoprotein (G-protein), and the protein matrix under this envelope consists of two proteins. A centrally located helical negative-sense single-stranded RNA is surrounded by a capsid of three proteins. Two of these show a combined RNA polymerase activity and are essential for virus replication. The virus attaches to the cell through the G-glycoprotein. Antibodies against this protein exhibit neutralizing activity, and studies of these antibodies have revealed several antigenic variants of rabies virus. It has not been established whether the antigenic differences are of importance for virulence or relevant design of vaccines. Viral infectivity is destroyed within 10 minutes at 60°C and by common disinfectants.

EPIDEMIOLOGY

Rabies occurs worldwide in non-domesticated mammals, which represent the reservoir of the virus. The natural susceptibility to rabies virus infection shows wide variation among species, but experimentally all mammals examined can be infected. In Europe foxes are the most common source of infection. Transmission of rabies from the reservoir to humans occurs mainly through dogs, in some cases through cats, cattle and bats. Rabies virus is present in the saliva, and humans contract the infection when bitten or licked by infected animals. Some animal species may carry the virus for long periods without exhibiting clinical manifestations; others may be clinically ill without excreting virus in the saliva. It is not clear if there is a reservoir of asymptomatic chronic carriers of the virus, for example small rodents. Most countries have a constantly infected stock of wild animals, and rabies frequently occurs among non-domesticated dogs in developing countries. In Europe, Scandinavia and the UK have been rabies-free for many years, and there are strict quarantine requirements for import of animals. Humans are not particularly susceptible to rabies, and human-to-human transmission is extremely rare. Although documented exposure is not infrequent, missionaries and health workers in developing countries very rarely contract rabies. Under special conditions infection through aerosols has occurred, for example in caves inhabited by bats and by exposure in the laboratory. Vaccination of domestic animals is an important control measure in countries where rabies occurs in the wild animal stock. In recent years extensive use of vaccine-containing baits has dramatically reduced rabies amongst wild animals in Europe.

THERAPY AND PROPHYLAXIS

The treatment after exposure is thorough washing of the actual site and administration of human specific **immunoglobulin** immediately followed by vaccination. Most modern rabies **vaccines** contain inactivated rabies virus originally grown in human diploid cells, Vero cells or chick embryo cells. These vaccines convey a high degree of immunity and are almost free from complications. The supply of vaccine is, however, relatively limited and the price is high. Pre-exposure prophylaxis is recommended for persons who are at high risk, particularly in connection with prolonged stay in enzootic areas where access to modern rabies vaccines may be difficult. Pre-exposure vaccination is given as three intramuscular (deltoid area) injections at intervals of 1 and 3 weeks, respectively. A similar intracutaneous regimen using 0.1 ml doses of high potency vaccines is cheaper and when correctly administered equally effective. Booster doses are recommended after one year and, when remaining in high risk areas, possibly every 5 years. The indications for use of vaccine and immunoglobulin **after exposure** should be established by experienced health authorities and, if indicated, implemented without delay. The WHO guidelines for postexposure immunization may be summarized as follows: Rinse immediately the exposed site thoroughly with soap and water, thereafter with 40–70% ethanol or iodine tincture. If the exposure is considered to be serious, 20 units/kg bodyweight of human specific antirabies immunoglobulin is given. If not available, highly purified horse antirabies immunoglobulin may be used. Half the dose is infiltrated at the location of the bite, the rest is injected intramuscularly into the deltoid area (or antero-lateral thigh of small children). At the same time a course of vaccination is started (5 doses on days 0, 3, 7, 14 and 30, respectively). If the risk of exposure is deemed to be small, vaccination is performed without giving immunoglobulin initially. Rabid animals usually die within a few days. It is therefore important to catch and isolate the suspect animal for observation. If the animal remains healthy 10 days after the person was exposed, the vaccination schedule can be changed to the pre-exposure schedule. Where indicated, antitetanus vaccination and/or antimicrobial treatment should be implemented.

LABORATORY DIAGNOSIS

The diagnosis of human or animal rabies has to be based on the following findings:

- Demonstration of virus antigens or nucleic acid from brain, spinal cord, salivary glands, saliva, cornea or skin by means of immunofluorescence or PCR, respectively.
- Postmortem demonstration of Negri bodies in brain tissue.
- Isolation of virus from brain tissue and/or saliva.

Antibodies cannot usually be demonstrated before clinically manifest disease, and in many cases serological tests are also negative throughout the clinical course.

WARTS AND ALL

38. HUMAN PAPILLOMAVIRUS (HPV)

Lat. *papilla* = nipple; Gr. *oma* = tumour.

T. Traavik

73 genotypes of HPV had been identified by 1994, but the number will most certainly increase. HPV infects mucosal and cutaneous epithelia. Many of the genotypes have been detected only in benignly proliferative processes, i.e. common warts, but an increasing number have been discovered in malignant tumours. Hence, the aetiological relationship between some HPV genotypes and carcinoma of the uterine cervix, as well as some other anogenital cancers, is well established. Evidence linking HPV to some cancers of the skin and the nasal/oral cavities is forthcoming.

TRANSMISSION/INCUBATION PERIOD

Infectious virus is shed from warts and transmitted by direct and indirect contact including sexual intercourse. Direct infection of basal layer cells in minor skin or mucosal lesions may play an important role. Anogenital HPV infections in early childhood can result from poor hygienic conditions, but may also indicate sexual abuse. Condylomatous lesions in children do, however, frequently contain HPV types regularly found at other cutaneous sites. Plantar warts are often spread in public baths. The incubation period is difficult to establish; 1–2 weeks has been suggested.

CLINICAL FEATURES

SYMPTOMS AND SIGNS

Common skin warts mostly affect children and young adults, with multiple lesions on hands and fingers. Flat skin warts tend to localize to arms, knees and the face. Deep plantar warts are usually single and are seen in children and adults. Warts may occur on the vocal cords (laryngeal papillomas).

Genital warts (condylomas), localized to cervix, vagina, vulva, penis and perineum, are transmitted by sexual contact, and appear as flat (condyloma planum) or exophytic (condyloma acuminatum) warts. In epidermodysplasia verruciformis, a rare inherited immunological disease, the warts may turn into invasive or *in situ* squamous-cell carcinomas upon exposure to sunlight. Juvenile laryngeal papillomas usually result from infection contracted at birth from genital warts in the mother.

Warts may give no symptoms aside from laryngeal irritation and the pressure pain associated with plantar warts. They are easily diagnosed clinically by their horny, papilliform appearance and small bleeding points upon scraping. An overwhelming amount of evidence is linking specific HPV genotypes to cancer of the cervix. Such 'high-risk' HPVs are associated with squamous intraepithelial neoplasias that are potentially precancerous. In the cervix this association concerns cervical intraepithelial neoplasia (CIN). CIN lesions are considered preneoplastic in that a small percentage will progress to cervical cancer. Altogether, approximately 85% of cervical cancers have been shown to contain DNA of one of the high-risk HPV types.

Differential diagnosis. Warts must be distinguished from hyperkeratotic plaques and squamous-cell carcinomas, as well as from *molluscum contagiosum* which is caused by a poxvirus.

CLINICAL COURSE

Skin warts and most condylomas heal spontaneously, but may persist for years. Cervical cancer is the second leading cause of death from cancer among women worldwide, with approximately 500,000 deaths annually.

THE VIRUS

Papillomavirus and *Polyomavirus* are the two genera that constitute the *Papovaviridae* family, but the *Papillomavirus* genus (Figure 38.1) has many distinctive properties. They are small, 'naked' double-stranded DNA viruses, which cannot be grown in cell culture. Replication takes place in the nucleus and only in differentiating cells such as the keratinocytes of the skin. Large amounts of virus are produced from skin warts, while complete virions are not

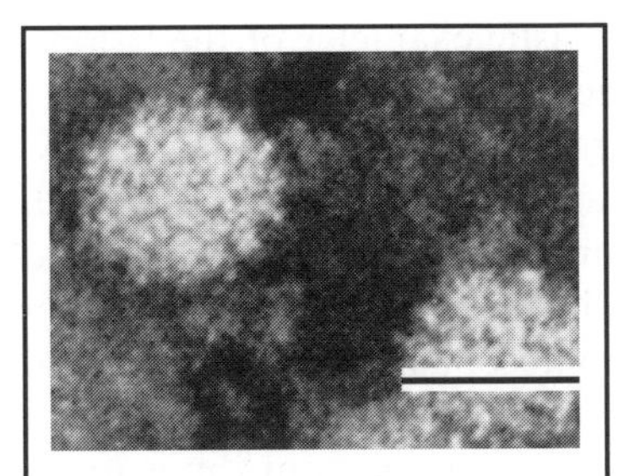

Figure 38.1 HUMAN PAPILLOMAVIRUS. Bar, 50 nm (Electron micrograph courtesy of G. Haukenes)

released from malignant lesions. The size of the double-stranded, circular HPV genome is a little less than 8 kbp. It codes for seven early, regulatory polypeptides (E1–E7) which are not included in the virions, and two late, structural proteins (L1–L2). Viral DNA replication and expression of the late proteins are restricted to differentiating cells. This block has prevented efforts to efficiently propagate HPV in culture. E1 and E2 are DNA-binding proteins that are involved directly in regulation of viral, probably also cellular, transcription and DNA replication. Similarly to the oncoproteins of other DNA tumour viruses, the E6 and E7 proteins interact with and change the activity of cell regulatory proteins that govern the admission into, as well as the speed of, cell division cycles and differentiation pathways.

Based on DNA sequence homologies, 73 genotypes of HPV had been identified in 1994, and each is associated with a specific clinical entity. At present several other types are awaiting final inclusion. A viral strain is considered new when the DNA sequences in specific regions of the viral genome (E6, E7 and L1) differ by more than 10%. For the majority of the HPV genotypes approved so far, the genomes have been completely sequenced, allowing the establishment of closely or more distantly related groups. This grouping reveals genetic distinctions between HPVs preferentially infecting mucosal and those infecting cutaneous locations, with the genotypes causing epidermodysplasia verruciformis forming a subgrouping of the latter. Many of the human genotypes are less related to each other than to papillomavirus from other animal species.

ONCOGENICITY AND ROLE OF HPV IN HUMAN CANCERS

Common warts and plantar warts never become malignant. Direct conversion of condylomas to carcinomas has been described anecdotally but is extremely rare. In addition to experimental evidence, large-scale epidemiological studies performed during recent years have established HPV infections as the major risk factor for cervical cancer. Skin cancers arising in immunosuppressed patients are also increasingly found to contain HPV DNA. Since the matter of causality is being elucidated, the focus is at present, and in the years to come, being shifted towards the mechanisms by which the interplay between certain HPV genotypes, the infected host cells, their environment and other factors may initiate, establish and maintain oncogenic processes in the cervix as well as other locations. HPVs associated with anogenital lesions have been divided into 'low-risk' types (6, 11, 34, 40, 42, 43) or 'high-risk' types (16, 18, 31, 33, 35,

39, 45, 51, 52, 54, 56, 58) based on the preneoplastic character of the lesions. Low-risk HPVs such as types 6 and 11 are generally associated with venereal warts or condylomas acuminata, which only rarely progress to malignancy. High-risk HPVs include types 16 and 18 and are associated with CIN. Both these HPVs were originally isolated from human cervical carcinomas, and approximately 70% of all cervical cancers contain either HPV16 or 18. In CIN lesions the HPV genomes are found extrachromosomally, while in cervical cancers viral DNA is found integrated. Integration may confer a relative growth or survival advantage to the host cell, and hence contribute to an oncogenic process. Continued expression of the E6 and E7 genes is, however, necessary for maintained proliferation of cervical carcinoma cells. The mechanisms by which the E6 and E7 proteins of high-risk HPVs contribute to cellular transformation and carcinogenesis are not known in detail but, similarly to oncoproteins of other DNA tumour viruses, these HPV proteins interact with and change the activities of cellular regulatory proteins. The best known examples are the complex formations with and functional neutralization of the tumour suppressor proteins p53 (E6) and pRB, the product of the retinoblastoma susceptibility gene (E7). The E6 and E7 proteins of low-risk HPVs are less transforming, and bind the tumour suppressor proteins with lower affinity than their counterparts from high-risk types.

EPIDEMIOLOGY

Papillomaviruses are widespread in many different animal species. In man, skin warts are easily spread among children, and the occurrence of genital warts is closely related to sexual activity. There are reports about earlier manifestation of cervical and penile cancers in geographical areas where onset of sexual activity is early. Other observations clearly show that promiscuity represents the prime risk factor for the development of cervical cancer.

Papillomaviruses are remarkably host-specific. Many animal species, including all the common domesticated ones, have their own papillomaviruses. In some instances HPVs are genetically more closely related to animal representatives than to other human types. In spite of this, there are no examples of papillomaviruses crossing species barriers.

Obviously, many HPV infections remain clinically inapparant. They may induce microfoci which are barely visible to the naked eye, but still contribute to the spread of these infections. The modes of transmission are still not fully understood. For some HPV types sexual activity may be the dominant mode, but additional routes, including perinatal and other forms of contact infection under poor hygienic conditions, may also prove important. Epidemiological studies have been difficult due to the lack of diagnostic methods to determine past exposure to specific HPV types.

IMMUNE RESPONSE

Antibodies directed against conformational epitopes on intact virions are evoked during natural infection. Low levels of such anti-HPV1 antibodies have been detected with a frequency of 10–100% in individuals with warts as well as in populations of unselected individuals. Responses to the major L1 and minor L2 capsid antigens of various HPV types have been reported, but do not seem to correlate with HPV-associated lesions or level of sexual activity. Serological responses against non-structural HPV proteins (E2, E4, E6, E7) have also been reported. Such antibodies seem to be virus-type and cancer-specific markers, but do not seem to have any clear prognostic value.

Cell-mediated immunity (CMI) seems to play the major role in combating HPV infection. Genetic, acquired or iatrogenic CMI deficiencies are accompanied by high frequencies of cutaneous or genital warts as well as anogenital neoplasias. Cytotoxic T-lymphocytes, mainly directed against the E6 and E7 proteins, seem to play a crucial role. Natural cytotoxicity may also be important for immunosurveillance. HPV-transformed cells have been lysed by NK cells and activated macrophages, and their proliferation is inhibited by TNF-α. HLA DQW3 is associated with a higher risk of developing cervical carcinoma, suggesting an immunogenetically controlled susceptibility.

THERAPY AND PROPHYLAXIS

Antigenic epitopes on the E6/E7 proteins of high-risk HPVs, and their interaction with CMI effector functions, are being extensively studied. The ultimate goal, a protective vaccine, does not seem to be within short-term reach.

Most skin warts regress spontaneously. Various local treatments have been attempted with uncertain effect (podophyllin, salicylic acid, glutaraldehyde, acetic acid and freezing). When the wart is troublesome, surgical removal may be necessary. Interferon has been used for treatment of laryngeal papillomas with favourable results. In approximately 50% of refractory cases, however, relapse is common. Education concerning prevention of sexually transmitted diseases should include appropriate counselling about HPV infections.

LABORATORY DIAGNOSIS

Papillomavirus particles are easily demonstrated by electron microscopy of skin wart scrapings, while there may be too few particles in condylomas to be detected by direct electron microscopy. HPV cannot be cultivated.

There are two clinical uses of HPV testing where the predictive values may be high enough to justify general use as an adjunct to cervical cytological screening: the clarification of equivocal and low-grade Pap smears, and the general screening in older women. HPV testing has generally been conducted by Southern blotting, PCR methods or commercially available dot–blot kits. Typing of high-risk and low-risk HPV strains is tedious with the two former methods, while the latter has so far been very selective and picks up only a few strains of each category. In addition to simpler typing procedures, quantitative data reflecting viral concentration have been warranted, since the viral load may correlate with the grade and progression of cervical pathology. For this purpose non-radioactive DNA detection methods like antibody-based hybrid capture and automated ELISA reading systems for detection of HPV PCR products are being developed. At the moment there are no diagnostically meaningful serological methods available, but antibodies to specific epitopes of HPV16 or 18 E6/E7 have recently been shown to be found more in cancer patients.

SERIOUS RELATIONS

39. HUMAN POLYOMAVIRUSES

Gr. *poly* = many, *oma* = tumour.

T. Traavik

Independent isolations of two strains of human polyomaviruses were achieved in 1971. The strains were given names after the initials of the patients from whom they were recovered, and have ever since been known as BK virus (BKV) and JC virus (JCV), respectively. Although clearly different species, the viruses are closely related to each other and to the macaque polyomavirus SV (simian virus) 40. BKV and JCV are widely distributed in human populations all over the world, and most people are infected before adolescence. BKV and JCV are so far the only human polyomaviruses characterized. JCV is gaining increasing clinical importance as the cause of PML (progressive multifocal leukoencephalopathy), a fatal degenerative brain disease affecting people with impaired cellular immunity. BKV has been associated with cases of urinary tract disease and respiratory symptoms, and may also be a more common multiorganic, opportunistic pathogen in immunosuppressed individuals than earlier realized. Both viruses have an oncogenic potential, but the clinical significance of this is still an open question. In 1994, strong indications for a polyomavirus as an aetiological cofactor for mesothelioma were published. This polyomavirus, which is still not finally identified, appears to act as a cofactor with asbestos in the development of this invariably fatal form of lung cancer, estimated to kill 80,000 people before the year 2015 in the USA alone.

TRANSMISSION/INCUBATION PERIOD

The routes of transmission and spread within individuals are virtually unknown for both BKV and JCV. Respiratory and/or oral transmission routes with viraemic spread of viruses to the kidneys and other organs are generally suspected. Primary infection is followed by persistence or latency. Latent infections may be reactivated by immunosuppression, some chronic diseases and pregnancy, even when high levels of serum antibodies are present.

SYMPTOMS, SIGNS AND CLINICAL COURSE OF PROGRESSIVE MULTIFOCAL LEUKOENCEPHALOPATHY (PML)

PML is an insidious, subacute disease with minor or no inflammatory reactions. Lesions are seen in multiple foci of demyelination. Microscopically the striking feature is oligodendrocytes with enlarged, deeply basophilic nuclei. The nuclei are filled with virions and viral proteins. The cerebrum is the part of the brain most often affected. Early manifestations include both motoric and mental disturbances as well as visual and speech impairment. Complete or partial paralysis, dementia and blindness may follow. The survival time of PML following diagnosis is often less than 6 months.

THE VIRUS

Polyomavirus forms one of the two genera within the family *Papovaviridae*, the other being *Papillomavirus*. The two genera are not in any way related, and ought to be reclassified accordingly. Recent efforts have been made to rename BKV and JCV as Polyomavirus hominis 1 and 2, respectively, but the new labels have not stuck so far.

Polyomaviruses have non-enveloped, icosahedral virions with a diameter of 38–44 nm which are resistant to ether, acid and heat treatment. The genome consists of a double-stranded, covalently closed circular DNA molecule of approximately 5 kbp, which is packed with host-cell histones as a minichromosome. The genome can be divided into three functional regions. The two protein-coding regions run in opposite directions on different strands. The early region is expressed before DNA replication, and codes for the tumour antigens T and t. The late region mainly codes for the three capsid proteins VP1–3. The two coding regions are separated by a non-coding control region (NCCR) containing the origin of replication and promotor/enhancer for both early and late transcription. The coding regions show high degrees of sequence homology (≥70%) between primate polyoma species, and near identity between different isolates of strains from the same species. The NCCR, however, is hypervariable between and within species, and affects host-cell tropism and oncogenic potential.

ONCOGENICITY AND ROLE IN HUMAN CANCERS

Fully permissive host cells are usually killed by polyomavirus infection. Infected cells that have the viral tumour antigen expressed, and survive, may become immortalized and malignantly transformed. Large T-antigen contributes to this by at least two sets of activities: as a host-cell transcription factor and by complexing with and inactivating the tumour suppressor p53 as well as the retinoblastoma gene product pRB, and probably also other host proteins that regulate cell cycle progression and growth. Small t-antigen also acts as a transactivating factor, most probably by complexing with and inactivating protein phosphatase 2A, a mechanism which keeps important intracellular signal transduction pathways active. The oncogenic potential of primate polyomaviruses is evident by infection of rodents, while in the authentic host systems the viruses seem to require cofactors to be transforming. Human polyomaviruses are certainly not *the* cause of any human cancer. To what extent some viral strains may be part of an arsenal of alternative initiating and maintaining aetiological factors behind some cancers of the brain, bones, pancreas, intestines and other organs, is still an open question.

IMMUNE RESPONSE

Most people have antibodies against BKV and JCV. High antibody titres may persist for life, possibly due to frequent reactivation of latent infections. Cellular immune responses are not well studied, but relative deficiencies in such responses may precede reactivations.

THERAPY AND PROPHYLAXIS

Specific treatment or vaccines are not available. Treatment of PML with nucleoside analogues and interferons has not been successful.

LABORATORY DIAGNOSIS

Sensitive ELISA tests for serological diagnosis and PCR-based methods for viral identification of human polyomavirus have been developed, and services are available from selected laboratories.

SLOWLY BUT SURELY PAST THE HOST BARRIER

40. SLOW VIRUSES

The term 'slow virus infection' was introduced by the late Icelandic veterinary surgeon Bjørn Sigurdsson in 1954.

G. Haukenes

Slow virus infections are progressive disorders with a long incubation period (years), proceeding slowly to death. The causative agents are either conventional viruses or so-called unconventional or atypical agents, recently also called prions.

CONVENTIONAL SLOW VIRUS INFECTIONS

SUBACUTE SCLEROSING PANENCEPHALITIS (SSPE)

SSPE is caused by measles virus which becomes reactivated in brain cells years after the primary acute measles infection. The majority of SSPE patients contracted their measles infection in early infancy, before 2 years of age, and the interval to development of SSPE is 5–10 years. Thus, in most cases the SSPE patients are children or teenagers. Boys are more commonly affected than girls.

The main **clinical manifestations** are intellectual deterioration and motor dysfunction. The disease progresses to death in the course of 4–12 months. The most significant diagnostic finding is the presence of oligoclonal IgG bands on agarose electrophoresis of concentrated cerebrospinal fluid (CSF). The bands represent antibodies to various measles virus components. The antibody titres to measles virus antigens in the CSF, examined by haemagglutination inhibition, complement fixation or ELISA tests are raised relative to those of serum. Normally the antibody concentration in CSF is 100–800 times lower than in the serum. In SSPE this ratio is dramatically reduced for measles antibodies, indicating intrathecal production of antibodies provided that a damaged blood–brain barrier can be excluded.

The **pathogenesis** of SSPE is not clear. It is not known in what form and where the measles virus or viral genetic material persists in the period from acute measles to the development of SSPE. Measles virus is an RNA virus without a reverse transcriptase or other enzymes required for integration of a proviral DNA, as is seen for retroviruses. SSPE virus is in some way defective as it cannot be cultivated from brain material unless a co-culture system with brain cells and a measles virus permissive cell line is established. During replication there is inadequate synthesis of the virion matrix protein which constitutes the inner part of the viral envelope. Complete virions are thus not released from the infected cells, and the virus is protected against the immune response during its passage from cell to cell. There is also experimental evidence for an antibody-mediated modulation of the viral antigens at the surface of the infected cells, resulting in immune complex formation and 'stripping' of antigens. In this way the cell will not become a target for killer cells. The main cell type infected in SSPE is the oligodendrocyte, a supporting cell which produces myelin, leading to demyelinization and neuron dysfunction. The yearly incidence of SSPE is 1–2 per million. After introduction of the live measles virus vaccine, the occurrence of SSPE has been reduced at least 10-fold.

A very rare SSPE-like disorder has been described to be caused by rubella virus. This **progressive rubella panencephalitis** develops 6 months to 4 years after the primary rubella infection, which in most of the few cases described was a congenital rubella.

PROGRESSIVE MULTIFOCAL LEUKOENCEPHALOPATHY (PML)

The causative agent of this disorder belongs to the polyomavirus group. Strain JC virus was isolated from brain tissue from PML patients by co-cultivation. These viruses are common as antibodies to them are found in most humans. The clinical manifestation of the primary infections is unknown. PML occurs only in patients who are immunosuppressed either from disease (including AIDS) or from treatment with immunosuppressive drugs. The diagnosis can be made by examination of affected brain tissue for intranuclear inclusion bodies. The inclusion bodies can be shown to contain aggregates of polyomavirus by electron microscopy.

Some other chronic persistent infections have many features in common with slow virus infections such as chronic active hepatitis B and AIDS/HIV infection. In animals the prototype slow virus infections are visna and maedi in sheep caused by a lentivirus (see Chapter 31).

UNCONVENTIONAL AGENTS/ATYPICAL VIRUSES/PRIONS

The first disease caused by an unconventional agent to be described was **scrapie** in sheep. Scrapie is an encephalopathy leading to ataxia and other motor dysfunctions. The incubation period is about 1–6 years. Most experimental studies on this group of unconventional agents have been made with scrapie transferred to mice in which the incubation period may be reduced to a few months. Scrapie occurs naturally in sheep, goats, cattle (bovine spongiform encephalopathy, BSE) and mink (transmissible mink encephalopathy) but can be transmitted to several other animals, including mice.

Kuru, a disease resembling scrapie in sheep, was described in humans by Zigas and Gajdusek in 1957. The disease had occurred for some decades in the small Fore tribe of Papua New Guinea and was named kuru by the natives. At the time kuru was described, it had attained epidemic proportions, being the most common cause of death in children and young women. Gajdusek showed that kuru was transmitted by an infectious agent. He inoculated brain material from kuru patients into chimpanzees, which later developed a similar disorder. From apes the agent has been further transmitted to several other animal species. Women and children contracted the infection when they prepared the brain material from deceased relatives for meals as part of the tribe's ritual cannibalism. When this cannibalism was abandoned and ceased in 1960, no new cases of kuru appeared which could not be ascribed to earlier cannibalism.

Creutzfeldt–Jakob disease (CJD) is a presenile dementia which is usually seen in persons between 45 and 60 years of age. Most cases develop spontaneously, but accidental transmission has been reported in connection with cornea transplantation, surgical operations using contaminated instruments, hypophyseal extract injections and application of improperly decontaminated EEG electrodes for examination of stereotaxis. The incidence of CJD varies considerably globally, but it is rare and a mean yearly incidence of 1–2 per million population has been suggested. The main symptoms are mental retardation and motor dysfunction, leading to death within a year.

A novel human prion disease, new variant CJD, was first documented in 1986 and shown to be caused by exposure to BSE. The number of new variant CJD cases now exceeds 100 in Europe, of which 99 are in the UK. The disease has stricken almost exclusively people below 55 years of age and differs clinically from sporadic CJD.

The **unconventional agents** isolated in scrapie, kuru and CJD seem to be largely similar with their host-species tropism. It has not been possible to cultivate or to isolate, purify and characterize the causative agent. It seems to have a size of 20–100 nm diameter and is exceptionally resistant to heat, proteolytic enzymes, irradiation and many disinfectants. Autoclaving combined with strong oxidizing agents, such as sodium hydroxide, destroys the infectivity. Because of the high resistance to irradiation it has been concluded that the agent contains either no or only a small segment of nucleic acid as seen in viroids in plants. Considerable interest has been focused on a unique protein called prion (proteinaceous infectious particle, PrP) which is regularly found in scrapie plaques and is regularly associated with infectivity. The protein is built up of units of MW 27,000–30,000, aggregated into amyloid-like plaques. An identical protein has been demonstrated in healthy individuals coded by a gene located on chromosome 20. In contrast to scrapie PrP, the normal PrP does not become folded into fibrils and is sensitive to proteases. The various infectious prions appear to be abnormal isoforms of the host encoded prion protein.

The **pathogenesis** of these encephalopathies is unknown. Histologically the affected areas in the brain show degeneration of neurons with vacuolization (spongiform encephalopathy), a reactive astrocytosis and amyloid-like plaques. There are no signs of inflammatory reaction or immune response, and interferon production is not induced.

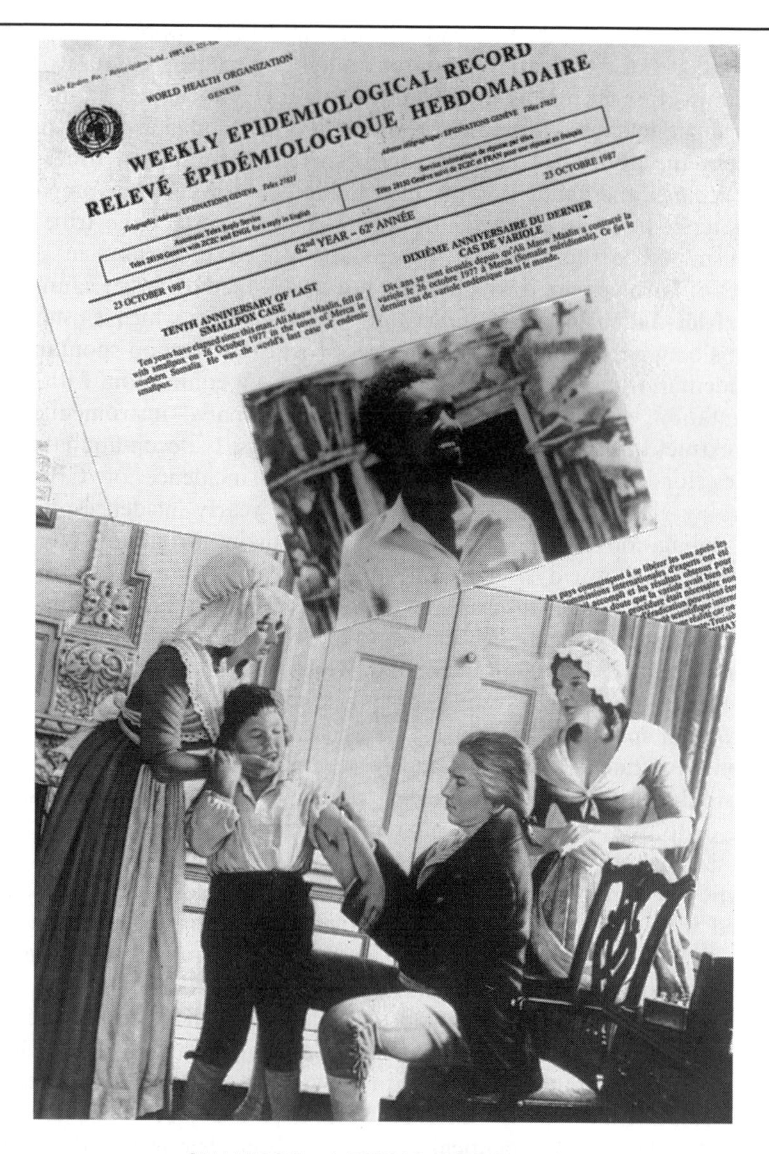

SMALLPOX—A THING OF THE PAST

Facsimile of the front page of *Weekly Epidemiological Record*; 62 (43), 321, 1987, reproduced by permission of the World Health Organization. Tenth Anniversary of the last smallpox case.

Edward Jenner, in his apartment in Chantry House, Berkeley, inserts exudate from a cowpox pustule on the hand of dairymaid Sarah Nelmes into scratches on the arm of 8-year-old James Phipps, 14 May 1796. (Painting by R. A. Thom, reproduced by permission of the Parke-Davis division of Warner-Lambert.)

41. POXVIRUSES

Variola from Lat. *varius* = spotted; 'smallpox' and Fr. *petite vérole* = small syphilis (syphilis being the great pox).

G. Haukenes

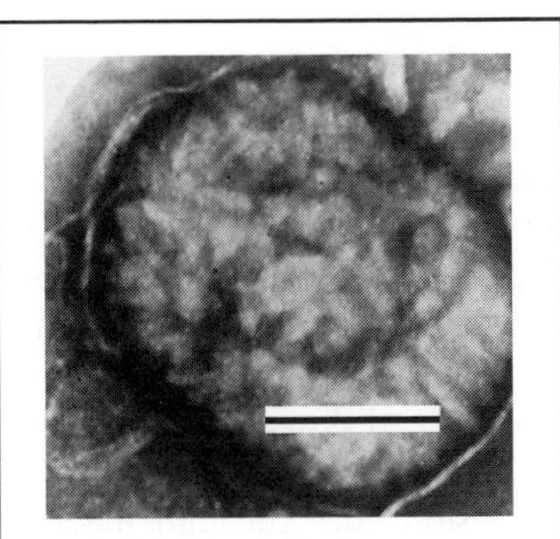

Figure 41.1 VACCINIA VIRUS. Bar, 100 nm (Electron micrograph courtesy of G. Haukenes)

Poxviruses (family *Poxviridae*) are large, brick-shaped or ovoid double-stranded DNA viruses of about 200–300 nm in diameter with a complex structure (see Figure 41.1). The poxviruses pathogenic for man include *Orthopoxvirus* (smallpox, vaccinia, monkeypox and cowpox viruses), *Parapoxvirus* (or/and pseudopoxviruses), *Molluscipoxvirus* and *Yatapoxvirus*. In humans they cause various exanthematous diseases with or without general symptoms and rare cases of tumour (Yaba monkey tumour virus).

ORTHOPOXVIRUS

Epidemics and endemics of smallpox have occurred for thousands of years in settlements in Asia and Africa. Also in more recent times smallpox has been a scourge to mankind. The last outbreak and the last case of smallpox transmitted from human to human were seen in Somalia in 1977. Two years later smallpox was declared to be eradicated throughout the world. Eradication of this disease had been achieved through the impressive World Health Organization (WHO) Smallpox Eradication Campaign. In the years to follow, vaccination against smallpox was gradually abandoned, first in civilians and later also in military and laboratory personnel.

Human orthopoxvirus infection is now restricted to some cases of monkeypox in Western and Central Africa and rare cases of cowpox in a few countries in Europe. The clinical picture of monkeypox is very similar to that of smallpox, while cowpox in humans is a mild disease if the affected individual has been vaccinated against smallpox. In the unvaccinated, serious complications similar to those seen after primary smallpox vaccination have been reported, such as generalized infection and severe infection of eczema. The host reservoir of cowpoxvirus is probably small rodents, and humans may contract the infection through contact with naturally infected animals such as cats and cattle. With increasing unvaccinated populations, the possibility of more severe human cases of cowpox in future cannot be excluded.

Orthopoxviruses grow in a wide range of tissue cultures. The various species are distinguished by typical lesions on the chorioallantoic membrane of embryonated hen's eggs or by gene technological methods. Recently considerable interest has been concentrated on the use of vaccinia virus (smallpox vaccine virus) as a carrier of genomic material from several viruses in the preparation of multivirus vaccines. Vaccinia virus is further attenuated and genomic material from other viruses is introduced into modified vaccinia virus genomes by the recombinant DNA technique.

PARAPOXVIRUSES

Parapoxviruses are widely distributed among different animal species, especially orfvirus and pseudopoxvirus (milkers' nodules, paravaccinia). The reservoir of orfvirus is sheep. Lambs develop granulomas (orf or contagious pustular dermatitis) around the mouth and can transmit the disease to humans by contact, usually to farmers, shepherds and veterinarians. The incubation period is 3–6 days. The human lesion appears on exposed skin areas, mostly on hands and arms and in the face. In rare cases secondary generalized eruptions are seen. The lesion goes through maculo-papular stages and develops in most cases into a granulomatous ulcer, 1–2 cm in diameter. It may be mistaken for a malignant growth or an anthrax pustule. There are usually no general symptoms. Super-infection with bacteria occurs. The lesion will regress completely after weeks or months. Secondary human cases are not seen. Farmers are familiar with clinical manifestations in both themselves and in sheep. The diagnosis can easily be verified by electron microscopic examination of scrapings or biopsy material from the lesion. Orfvirus is ovoid and elongated with a characteristic criss-crossing pattern on the surface (see Figure 41.2). There are no cell culture or antibody assays available for routine diagnostic use. A live vaccine for sheep has been used in some countries.

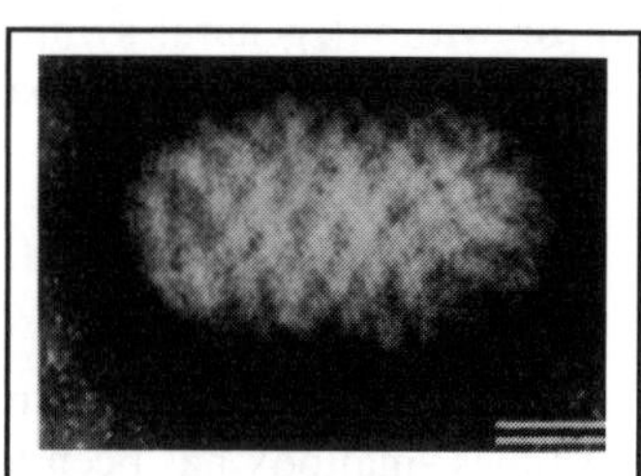

Figure 41.2 ORFVIRUS. Bar, 50 nm (Electron micrograph courtesy of G. Haukenes)

UNCLASSIFIED POXVIRUSES

Molluscum contagiosum virus affects only humans. It is transmitted by contact and occurs especially under poor hygienic conditions. The incubation period varies from 2 to 5 weeks. Most cases are seen in children or in young adults who have contracted the disease by sexual contact. Molluscum lesions are smooth, discrete skin nodules 1–5 mm in diameter, often with an umbilicated centre. The diagnosis is mostly made clinically but can be verified by electron microscopic examination of scrapings. The virus has not been cultivated, and there are no tests for antibody. The condition is mild, but in some cases the lesion may be long-lasting and has to be removed for cosmetic reasons. There is no vaccine against the disease.

YATAPOXVIRUSES

Tanapoxvirus (first recognized in the Tana River of Kenya) causes a mild zoonosis in monkeys and can occasionally be transmitted to man, presenting as a mild exanthematous disease with fever. Rare cases of Yaba monkey tumour (epidermal histiocytoma) virus transmission to man have been seen.

DON'T WORRY—IT'S ONLY A VIRUS!

42. CLINICAL SYNDROMES

G. Haukenes and J. R. Pattison

Clinical syndrome	Common aetiological agents (including common non-viral causes)
Adult T-cell leukaemia	HTLV-1
AIDS	HIV-1 and HIV-2
Arteritis, *see* Polyarteritis	
Arthritis/arthralgia/arthropathy	Seen in various virus infections (rubella, rubella vaccination, parvovirus B19, mumps, alphaviruses) and in chlamydial (Reiter's syndrome) and bacterial infections (*Haemophilus influenzae*, campylobacter, gonococci, streptococci group A)
Balanitis	Herpes simplex virus
Bronchiolitis	Respiratory syncytial virus Parainfluenzaviruses, adenoviruses
Cervicitis	Herpes simplex virus (*Chlamydia trachomatis*)
Chronic fatigue syndrome, *see* Postviral chronic fatigue syndrome	
Condyloma acuminatum	Human papillomaviruses
Congenital disease (*see also* Perinatal generalized infection)	Rubella virus Cytomegalovirus Herpes simplex virus (*Toxoplasma gondii* and *Treponema pallidum*)
Conjunctivitis (*see also* Keratoconjunctivitis)	Adenoviruses Enterovirus 70 (*Chlamydia trachomatis*)
Contagious pustular dermatitis	Orf virus
Creutzfeldt–Jakob disease	Prion

Continued

Clinical syndrome	Common aetiological agents (including common non-viral causes)
Croup	Parainfluenzaviruses Respiratory syncytial virus Influenzaviruses Measles virus (Note: *Haemophilus influenzae* epiglottitis)
Cystitis	Adenoviruses, BK virus
Diarrhoea, *see* Gastroenteritis	
Ecthyma contagiosum, *see* Contagious pustular dermatitis	
Encephalitis/meningitis	Herpes simplex virus Enteroviruses Varicella-zoster virus Alphaviruses Flaviviruses Bunyaviruses Rabies virus Mumps virus Measles virus (SSPE) Encephalitis lethargica—von Economo—seen 1914–1925 (postinfectious: measles, rubella, varicella)
Epidemic myalgia/pleurodynia	Coxsackie B virus
Epididymitis	Mumps virus (Gonococci and *Chlamydia trachomatis*)
Erythema infectiosum	Parvovirus B19
Exanthema subitum (Roseola infantum)	Human herpesviruses 6 and 7
Facial nerve palsy	Varicella-zoster virus Herpes simplex virus (In most cases the aetiology cannot be established)
Fever—often with a non-specific rash	Enteroviruses (especially echovirus) Adenoviruses Epstein–Barr virus Cytomegalovirus
Gastroenteritis	Rotavirus Adenoviruses Norwalk agent

Clinical syndrome	Common aetiological agents (including common non-viral causes)
	Small round viruses Caliciviruses Astrovirus Coronaviruses (*Campylobacter*, *Escherichia coli*, *Salmonella*, etc.)
Gingivostomatitis	Herpes simplex virus Coxsackie A virus (as herpangina) (Anaerobic bacteria)
Glomerulonephritis	Hepatitis B virus Cytomegalovirus (Streptococci group A)
Guillain–Barré syndrome	Considered to be an immune complex disease secondary to infection, e.g. cytomegalovirus, Epstein–Barr virus (*Mycoplasma pneumoniae*, etc.)
Haemorrhagic fever	Alphaviruses, arenaviruses, bunyaviruses, filoviruses, flaviviruses
Haemorrhagic fever with renal syndrome, *see* Nephropathia epidemica	
Hand, foot and mouth disease	Coxsackie A viruses Other enteroviruses (Note: Foot and mouth disease in cattle may very rarely be transmitted to man)
Hepatitis, acute	Hepatitis A virus Hepatitis B virus Hepatitis C virus Hepatitis D virus Hepatitis E virus Hepatitis G virus Hepatitis is also seen in yellow fever, CMV and Epstein–Barr virus infections
Hepatitis, chronic	Hepatitis B virus Hepatitis C virus Hepatitis D virus
Herpangina	Coxsackie A viruses
Infectious mononucleosis	Epstein–Barr virus Cytomegalovirus (*Toxoplasma gondii*)

Continued

Clinical syndrome	Common aetiological agents (including common non-viral causes)
Influenza	Influenza A and B viruses
Keratitis, keratoconjunctivitis	Herpes simplex virus Adenoviruses Enterovirus 70 Varicella-zoster virus
Meningitis (*see also* Encephalitis)	Enteroviruses Mumps virus Adenoviruses Herpes simplex virus Lymphocytic choriomeningitis virus Epstein–Barr virus (bacteria, fungi, parasites)
Molluscum contagiosum	Molluscum contagiosum virus
Mononucleosis, *see* Infectious mononucleosis	
Myalgia (epidemic), *see* Epidemic myalgia	
Myocarditis	Coxsackie B viruses
Nephropathia epidemica	Puumala virus
Ockelbo disease	Alphavirus (Sindbis virus-related)
Orchitis	Mumps virus Coxsackie B virus (Gonococci, *Mycobacterium tuberculosis*)
Pancreatitis	Mumps virus Coxsackie B viruses
Pareses	Polioviruses Other enteroviruses Encephalitis viruses Rabies virus
Pericarditis	Coxsackie B viruses
Perinatal generalized infection (*see also* Congenital disease)	Herpes simplex virus Coxsackie B viruses Varicella-zoster virus (Streptococci group B, *Escherichia coli*, *Listeria*)
Pharyngitis, *see* Tonsillitis and Upper respiratory infection	

Clinical syndrome	Common aetiological agents (including common non-viral causes)
Pleurodynia, *see* Epidemic myalgia	
Pneumonia	Influenzaviruses Adenoviruses Respiratory syncytial virus Parainfluenzaviruses Enteroviruses Varicella-zoster virus Cytomegalovirus (*Mycoplasma pneumoniae*, *Chlamydia trachomatis*, *pneumoniae* and *psittaci*, *Coxiella burnetii*, *Legionella pneumophila*, pyogenic bacteria)
Polyarteritis nodosa	Hepatitis B virus (immune complex) Hepatitis C virus
Postviral chronic fatigue syndrome	Cause unknown. Some cases have been associated with persistent coxsackie B virus infection and EBV has been implicated in some cases
Progressive multifocal leukoencephalopathy	Polyomavirus
Rhinitis, *see* Upper respiratory infection	
Stomatitis, *see* Gingivostomatitis	
Tonsillitis	Epstein–Barr virus Adenoviruses (Streptococci group A, *Haemophilus influenzae*)
Upper respiratory infection	Rhinoviruses Coronaviruses Adenoviruses Influenzaviruses Parainfluenzaviruses Respiratory syncytial virus Enteroviruses
Urethritis	Herpes simplex virus (Gonococci, *Chlamydia trachomatis*)
Vaginitis/vulvitis	Herpes simplex virus (Gonococci)
Warts	Human papillomaviruses

Index